Technological Approaches to Human Performance Enhancement

MARJORY S. BLUMENTHAL, ALISON K. HOTTES, CHRISTY FORAN, MARY LEE

Prepared for the Department of Defense
Approved for public release; distribution unlimited

NATIONAL DEFENSE RESEARCH INSTITUTE

For more information on this publication, visit **www.rand.org/t/RRA1482-2**.

About RAND

The RAND Corporation is a research organization that develops solutions to public policy challenges to help make communities throughout the world safer and more secure, healthier and more prosperous. RAND is nonprofit, nonpartisan, and committed to the public interest. To learn more about RAND, visit www.rand.org.

Research Integrity

Our mission to help improve policy and decisionmaking through research and analysis is enabled through our core values of quality and objectivity and our unwavering commitment to the highest level of integrity and ethical behavior. To help ensure our research and analysis are rigorous, objective, and nonpartisan, we subject our research publications to a robust and exacting quality-assurance process; avoid both the appearance and reality of financial and other conflicts of interest through staff training, project screening, and a policy of mandatory disclosure; and pursue transparency in our research engagements through our commitment to the open publication of our research findings and recommendations, disclosure of the source of funding of published research, and policies to ensure intellectual independence. For more information, visit www.rand.org/about/principles.

RAND's publications do not necessarily reflect the opinions of its research clients and sponsors.

Published by the RAND Corporation, Santa Monica, Calif.
© 2021 RAND Corporation
RAND® is a registered trademark.

Library of Congress Cataloging-in-Publication Data is available for this publication.

ISBN: 978-1-9774-0804-4

About This Research Report

This report is one of a set of focused, short analyses that examine critical emerging technologies and provide perspective on their development around the world. The analyses cover quantum technology, technologies for enhancing human performance, semiconductor technology, the intersection of artificial intelligence and cybersecurity, deepfake generation and detection, and the use of patent data to assist in understanding global trends in emerging technologies. This report focuses on human performance enhancement.

The short technology analyses draw from open sources to provide a snapshot in time of the status of the technology; its plausible evolution; and the thought leaders, firms, institutes, and countries that are working on it (with particular interest in the prospects of potential adversaries of the United States). The sponsor emphasized the importance of examining commercial technologies; that examination is included in this report, along with an examination of government technology initiatives that might spur commercial offshoots. These short, focused assessments have two objectives. The first objective is to provide open source background for U.S. Department of Defense and Intelligence Community officials on commercially developed technologies that could have an impact on future military and intelligence operations. The second objective is to provide background for the multimethod analysis approach (documented as part of this set of reports) using a combination of data on patent filings, citations, scientific collaborations, and organizational capacities. Defense and Intelligence Community officials will benefit from this multimethod approach as one source of validated data to inform decisionmaking about research-and-development investments, collection priorities, collection targeting, foreign material acquisition, and needs for scientific talent.

The research reported here was completed in 2020 and underwent security review with the sponsor and the Defense Office of Prepublication and Security Review before public release.

RAND National Security Research Division

This research was sponsored by the U.S. Department of Defense and conducted within the Cyber and Intelligence Policy Center of the RAND National Security Research Division (NSRD), which operates the National Defense Research Institute (NDRI), a federally funded research and development center sponsored by the Office of the Secretary of Defense, the Joint Staff, the Unified Combatant Commands, the Navy, the Marine Corps, the defense agencies, and the defense intelligence enterprise.

For more information on the RAND Cyber and Intelligence Policy Center, see www.rand.org/nsrd/intel or contact the director (contact information is provided on the webpage).

Acknowledgments

The research team is grateful for the support of the U.S. Department of Defense and its encouragement to explore the set of complementary emerging technologies addressed in this report. The team appreciates the constructive feedback provided by Ritika Chaturvedi and Jonathan Moreno, who reviewed an earlier draft. Responsibility for the work remains with the team.

Summary

Multiple complementary advances in technology can enhance the performance of people engaged in defense activities—on behalf of the United States or its adversaries. These modalities of human performance enhancement (HPE) can be grouped into three principal categories: gene editing, applications of artificial intelligence (AI), and networked technologies that are wearable or even implantable (the so-called Internet of Bodies [IoB]). Drawing primarily from a literature review, this report offers an examination of these approaches to HPE.

It might take at least ten years for gene editing to be a reliable HPE modality. Motivated originally by the potential to avoid or alleviate disease, gene editing could foster specific physical attributes (e.g., ability to cope with low oxygen levels) that could aid warfighters. Safety and ethical concerns have been prominent for gene editing, but China is known to be investing heavily in related biotechnologies for reasons of national security and prestige.

HPE applications of AI are not likely to have an impact in the next five years, but this category of implantables promises to improve decisionmaking (through cognitive prostheses) and human-system teaming (not only through better system design but through implantable brain-computer interfaces). Safety and ethical challenges might slow development, but both Russia and China have been active in developing these technologies.

Of the three modalities, the IoB is farthest along in development. Some applications, associated with mobile health and fitness devices worn by users, are already in use. Over the next three to five years, the variety of wearables will grow, and there will be progress in technologies that can be implanted. Although today's technologies focus on communicating with a central entity (such as a software provider), potential will also grow for peer-to-peer networking.

Although the concept of HPE can evoke science fiction, all three of these modalities benefit from many years of attention from the research community, traditional scientific funders, and—at least as important—commercial interests, particularly over the past few years. Biotechnology, AI, and the IoB have all attracted considerable investment, which suggests both a push to realize benefit from those investments and an expectation that some benefits will emerge sooner rather than later. As shown in the open source literature review documented in this report, the potential for these technologies to both protect people and make them more effective in warfighting situations has attracted the interest and investment of the U.S. and Chinese militaries, with more-targeted investment from Russia.

Contents

Figures and Tables

Figures

Tables

CHAPTER ONE

Human Performance Enhancement: A Multimodal Process

Several methods of enhancing human performance might have implications for military and intelligence operations. This overview introduces three different paths to human-performance enhancement (HPE) that are addressed more fully in subsequent chapters: (1) genetic modifications, (2) artificial intelligence (AI), and (3) "Internet of Bodies" (IoB) approaches in which devices that are implanted in or worn by a human are connected to a network. Other approaches, such as pharmacology, are not addressed. The content of this report is drawn primarily from reviews of the literature, augmented by researcher expertise in associated areas.

Why Care About HPE?

Most of the research into HPE aims to improve human health and well-being—to understand and avoid or mitigate known sources of morbidity or impairment and to improve quality of life. Genetic mutations, for example, are associated with diseases and have been targeted by gene therapies.[1] From a military and intelligence operations perspective, however, HPE—by offering the potential to increase strength, speed, endurance, intelligence, and tolerance of extreme environments and to reduce sleep needs and reaction times—could aid in the development of better operators. These same developments also might create risks that could be exploited by adversaries.

---

[1] Genetic manipulation can also be used on plants and animals, contributing to improvements in the food supply that can affect people.

Thus, HPE can be used to assist healthy individuals, regardless of whether the underlying techniques were developed as therapies for people with disease or impairment.

The possible long-term applications of current research into gene editing, the genetic basis of human traits, cellular processes, and human physiology illustrate the potential implications. The 2019 Nobel Prize in Physiology and Medicine recognized work to understand how cells sense and adapt to the lack of oxygen (hypoxia).[2] This fundamental knowledge might one day contribute to HPE in oxygen-limited environments, such as high-altitude regions.

In the next few years, HPE could help military service and intelligence analysts through the use of multiple techniques to connect technology to human beings. Some of these techniques are being explored or used today. Technological devices can be worn external to the body (e.g., prosthetics, exoskeletons) or internally implanted.

Both of those capabilities, if designed to be networked, can be considered part of the IoB, an aspect of the Internet of Things (IoT) that focuses on devices worn on or taken into the body that communicate information about micro to macro levels of human performance. Work to date focuses on communications between individuals and a service provider, but it is not difficult to imagine broader networking (e.g., among people directly) and then consider its HPE potential.

A related kind of advance is the evolution of computer systems that use AI to support human decisionmaking. This can refer to what are sometimes called *cognitive prostheses*; it can also refer to improvements in support for human-machine teaming or collaboration. Goals include fluid cooperation and shared control between a human and a (computer-based) machine. The increasing speed and sophistication of systems, however, make it increasingly challenging to maintain a human in (or even on) the loop.

What Is the Time Horizon for HPE?

The most fundamental HPE acts at the gene level to change the attributes of an individual person (the phenotype). The full impact will be realized only in

---

[2] Nobel Foundation, "The Nobel Prize in Physiology or Medicine 2019," webpage, 2019.

the long term (more than ten years from now). In the nearer term, studies will continue to illuminate the connections between genotypes and phenotypes.

The science and the biotechnology for gene editing are both immature. Fueled by the steady growth in genomic data, efforts are being made to identify and make genetic changes (in humans or other species) that will cause a desired affect (e.g., reduced symptoms in cystic fibrosis patients, tumor shrinkage). Currently, approved therapeutic uses of gene editing modify only *somatic cells*—cells that will not be passed to future generations.

Genetic changes to human germline cells have the potential to affect future generations and thus prompt additional safety and ethical considerations. The focus of this report is on the technical aspects of HPE that might be relevant for military and intelligence operations; readers interested in the ethical implications of HPE might wish to consult such sources as *Emerging Cognitive Neuroscience and Related Technologies* from the National Academies of Sciences, Engineering, and Medicine.[3]

In additional to ethical concerns, the controversy over a Chinese scientist claiming to have created "CRISPR babies" who lack a gene associated with HIV also reflected the scientific community's concerns that (1) current gene editing techniques create unintended modifications (i.e., off-target changes), and (2) genetic changes can cause unintended physiological consequences.[4] Multiple genes interact to shape traits, and many genes influence multiple traits. Much work on secondary and tertiary effects is needed to reduce unintended consequences of genetic modifications intended for HPE. Thus, advances in this domain of HPE will be driven by technical challenges, ethical concerns, and policy and legal developments, all of which will influence the time needed for development.

Cognitive prostheses and other technologies that involve reading brain function might be available in the medium term (five to ten years). Perfecting the brain-computer interface (BCI) is a goal. As will be discussed

---

[3] National Research Council, *Emerging Cognitive Neuroscience and Related Technologies*, Washington, D.C.: National Academies Press, 2008.

[4] Jon Cohen, "The Untold Story of the 'Circle of Trust' Behind the World's First Gene-Edited Babies," *Science*, August 1, 2019. *CRISPR* (clustered regularly interspaced short palindromic repeats) refers to a family of DNA sequences. CRISPR technology is a method for making precise gene edits.

in Chapter Three, the prevailing technologies present portability and precision challenges.[5]

Other approaches to HPE can be used to varying degrees now or will be available soon. These are generally designed to be used by (or on) a person as he or she is. The IoB provides several illustrations that emerge from consumer and information technology and from medical applications. Because these applications involve supplementing or complementing existing human performance, they could be summarized as the *cyborg* approach to HPE—the augmentation of human performance by adding something to the human body.

Who Is Working on HPE?

The chapters that follow this overview discuss research, development, and implementation of a variety of technologies that tend to be anchored in specific contexts—cancer research, diabetes or seizure management, decision support, and physical performance.

HPE activity in all domains can be found in the three major powers—the United States, China, and Russia. It can also be found more selectively in Canada and among countries across Europe (notably the United Kingdom, France, and the countries of Scandinavia) and Asia (notably Japan and South Korea). This dispersion reflects the distribution of biomedical research and of information technology markets.

The biomedical strengths of the United States are well known. At the same time, U.S. strengths in information technology and in medical devices feed developments in the IoB.

China has targeted biotechnology, AI, and the IoT for growth. At least as important, China has become a global genetic and genomic data powerhouse—partly because it is home to a world-leading gene sequencing company (BGI Technology Company) and other big processors of genetic data (WuXi Nextcode Genomics and WeGene [Chinese population focus])

---

[5] These technologies include electroencephalography (EEG), functional magnetic resonance imaging (fMRI), magnetoencephalography (MEG), and positron emission tomography (PET).

and partly because access to large stocks of genetic data is comparatively easy as a result of government demands for personal information and, apparently, fewer concerns about privacy among the population of China than among those in Western nations. Furthermore, although the United States still leads China in the life sciences, China is steadily progressing down the path from copier to innovator.[6] Pursuit of HPE—notably military applications of neuroscience, preparation for fighting in virtual domains, and optimizing the integration of people and computer-based systems (especially those using AI)—is consistent with China's focus on "intelligentized" approaches to conflict, although the potential for gene editing to support ethnic attacks, albeit long term, has been noted by the People's Liberation Army.[7]

What Are Key Uncertainties?

Perhaps the biggest and most obvious uncertainty relates to differences in rules of engagement for HPE—beginning with norms and ethics. Normative and ethical stances vary across countries; they particularly shape the kind of research supported by governments.[8] Applicable international frameworks are both uncertain and evolving. U.S. scientists and officials have expressed concern that China, especially, might adopt a regulatory system with substantial differences from systems used in other countries.[9] The decision of a Chinese court to impose prison sentences on researchers who participated in the conception of the first genetically engineered

[6] Robert D. Atkinson and Caleb Foote, *Is China Catching Up to the United States in Innovation?* Washington, D.C.: Information Technology & Innovation Foundation: April 8, 2019.

[7] Elsa B. Kania, "Minds at War: China's Pursuit of Military Advantage Through Cognitive Science and Biotechnology," *PRISM*, Vol. 3, No. 8, 2020.

[8] Cortney Weinbaum, Eric Landree, Marjory S. Blumenthal, Tepring Piquado, Carlos Ignacio, and Gutierrez Gaviria, *Ethics in Scientific Research: An Examination of Ethical Principles and Emerging Topics*, Santa Monica, Calif.: RAND Corporation, RR-2912-IARPA, 2019.

[9] Weinbaum et al., 2019.

babies somewhat reduces these concerns, however.[10] Additionally, the International Commission on the Clinical Use of Human Germline Genome Editing (which was convened by the U.S. National Academy of Medicine, the U.S. National Academy of Sciences, and the United Kingdom's Royal Society in order to develop a framework for accessing clinical applications of human germline editing) contains two members from the Chinese Academy of Sciences.[11] Societal decisions about the appropriate use of genetic modifications should affect the types of genetic-based HPE that eventually become available. Additionally, the time it takes societies to work through the associated ethical and safety considerations will affect when different types of genetic-based HPE are first offered.

The lack of understanding regarding who will do what also contributes to uncertain time frames. The private and public sectors play different roles in different countries, and support for dual use innovation differs accordingly. Moreover, China has a strategy to participate in the leadership of international standards-setting related to information technology, which can affect many tools associated with HPE.[12]

Previously, open publication and international exchanges of ideas have contributed to advances in fundamental research. Reports of China using its Thousand Talents program and other approaches (such as encouraging violations of confidentiality in the peer-review process of U.S. grant-funding agencies to unfairly exploit basic research funded by the U.S. government)

---

[10] David Cyranoski, "What CRISPR-Baby Prison Sentences Mean for Research: Chinese Court Sends Strong Signal by Punishing He Jiankui and Two Colleagues," *Nature*, Vol. 577, 2020. Concerns about research ethics in China reflect a variety of reported practices; for example, see Amy Quin, "Fraud Scandals Sap China's Dream of Becoming a Science Superpower," *New York Times*, October 13, 2017.

[11] National Academies of Sciences, Engineering, and Medicine, "International Commission on the Clinical Use of Human Germline Genome Editing," webpage, undated.

[12] Dan Breznitz and Michael Murphree, *The Rise of China in Technology Standards: New Norms in Old Institutions*, research report prepared on behalf of the U.S.-China Economic and Security Review Commission, January 16, 2013; U.S.-China Economic and Security Review Commission, "Section 2: Emerging Technologies and Military-Civil Fusion: Artificial Intelligence, New Materials, and New Energy," in *2019 Report to Congress*, Washington, D.C.: U.S. Government Publishing Office, November 2019.

have attracted congressional scrutiny.[13] If the United States enacts policies to restrict the use of U.S. research, then international research collaboration networks could restructure and create difficult-to-predict implications for the rate of HPE progress and the eventual leaders in the field.

Finally, uncertainties about the possibilities for and consequences of device malfunction, device manipulation, and data exploitation make it difficult to understand the full risk picture.

---

[13] Brendan O'Malley, "China Is 'Systematically Stealing US Research'—Senate," *University World News*, November 22, 2019.

CHAPTER TWO

Potential for Using Genetic Modifications to Increase Human Performance

Advances in the fields of biology and medicine are rapidly increasing the availability of gene therapies to treat human diseases caused by genetic mutations. Some therapies under development treat inherited genetic diseases, such as cystic fibrosis and sickle cell disease,[1] for which the causative mutations are present in each of a patient's cells at birth. Other therapies target cancers that result from mutations accumulated in a subset of a patient's cells during his or her life.[2]

In addition to treating diseases through gene therapy, genetic modification techniques also have the potential to enhance normal humans. For example, some Duchenne muscular dystrophy gene therapy treatments could increase muscle mass in disease-free humans and thus be used for

---

[1] Eric W. F. W. Alton, David K. Armstrong, Deborah Ashby, Katie J. Bayfield, Diana Bilton, Emily V. Bloomfield, A. Christopher Boyd, June Brand, Ruaridh Buchan, Roberto Calcedo, et al., "Repeated Nebulisation of Non-Viral CFTR Gene Therapy in Patients with Cystic Fibrosis: A Randomised, Double-Blind, Placebo-Controlled, Phase 2b Trial," *Lancet Respiratory Medicine,* Vol. 3, No. 9, September 2015; Jean-Antoine Ribeil, Salima Hacein-Bey-Abina, Emmanuel Payen, Alessandra Magnani, Michaela Semeraro, Elisa Magrin, Laure Caccavelli, Benedicte Neven, Philippe Bourget, Wassim El Nemer, et al., "Gene Therapy in a Patient with Sickle Cell Disease," *New England Journal of Medicine,* Vol. 376, No. 9, February 28, 2017.

[2] U.S. Food and Drug Administration (FDA), "FDA Approval Brings First Gene Therapy to the United States," news release, August 30, 2017.

HPE.[3] Multiple academic authors have speculated about the potential for genetic modification to (1) make humans stronger, more intelligent, or more adapted to extreme environments and to (2) provide new capabilities (such as infrared vision)—applications with potential implications for military and intelligence operations.[4]

Identifying Genetic Changes with the Potential to Improve Human Performance

Improvements to human performance can be grouped into two categories: (1) improvements that provide a capability beyond that exhibited within the *natural* (i.e., nongenetically modified) human population, and (2) improvements that increase an individual's capability, but only to a level that still falls within the variation present in the natural human population.[5] Adding reptilian genes that provide the ability to see in infrared is an example of the first category.[6] Increasing an average runner's endurance to the level of an elite marathoner is an example of the second category.

To identify genetic changes with the potential to increase human performance, researchers can consider (1) the human population, (2) other species, and (3) novel genes designed de novo in a laboratory. Consistent with this report's emphasis on the near term, the focus of discussion will be on the first category: identifying changes within the human population with the potential to increase the capability of a human beyond his or her natural endowment of the capability.

---

[3] Ellen Wright Clayton, "A Genetically Augmented Future," *Nature*, Vol. 564, No. 7735, December 13, 2018.

[4] Braden Allenby, "Designer Warriors: Altering Conflict—and Humanity Itself?" *Bulletin of the Atomic Scientists*, Vol. 74, No. 6, 2018; Marsha Greene and Zubin Master, "Ethical Issues of Using CRISPR Technologies for Research on Military Enhancement," *Journal of Bioethical Inquiry*, Vol. 15, No. 3, 2018.

[5] National Academies of Sciences, Engineering, and Medicine, *Human Genome Editing: Science, Ethics, and Governance*, Washington, D.C.: National Academies Press, 2017.

[6] Allenby, 2018; Greene and Master, 2018.

The human population exhibits many phenotypes. For example, individuals vary in height, strength, speed, sleep needs, longevity, and susceptibility to diseases, with the relative contributions of genetics and the environment varying by trait (i.e., nature vs. nurture). Heritability accounts for the majority of the variation in such traits as maximum heart rate and aerobic capability that contribute to elite athletic performance.[7] Nonetheless, even monozygotic twins, who have identical genomes, are not actually identical.[8] The relative contributions of genetics and the environment to particular traits is an active area of research. For many traits, genetics and environmental contributions likely interact in complicated, nonadditive ways.[9]

Before the human genome had been sequenced, human genetic research focused on finding the genetic basis of phenotypes, such as Huntington's disease, that were caused by a single gene.[10] More than 5,000 phenotypes have been associated with a single gene.[11] Although the incidence of most genetic diseases is low, so many exist that approximately 1 in 17 people is affected.[12] The publication of the 3-billion base pair human genome sequence in 2001 enabled new and powerful approaches, such as genome-

---

[7] Evelina Georgiades, Vassilis Klissouras, Jamie Baulch, Guan Wang, and Yannis Pitsiladis, "Why Nature Prevails over Nurture in the Making of the Elite Athlete," *BMC Genomics*, Vol. 18, Suppl. 8, 2017.

[8] For a meta-analysis of twin studies, see Tinca J. C. Polderman, Beben Benyamin, Christiaan A de Leeuw, Patrick F Sullivan, Arjen van Bochoven, Peter M Visscher, and Danielle Posthuma, "Meta-Analysis of the Heritability of Human Traits Based on Fifty Years of Twin Studies," *Nature Genetics*, Vol. 47, No. 7, 2015.

[9] Bruno Sauce and Louis D. Matzel, "The Paradox of Intelligence: Heritability and Malleability Coexist in Hidden Gene-Environment Interplay," *Psychological Bulletin*, Vol. 144, No. 1, 2018. Because of the contributions of nongenetic factors, the approaches discussed in this chapter for understanding the genetic contributions to a phenotype only address a portion of the overall variability within the human population.

[10] Marcy E. MacDonald, Christine M. Ambrose, Mabel P. Duyao, Richard H. Myers, Carol Lin, Lakshmi Srinidhi, Glenn Barnes, Sherryl A. Taylor, Marianne James, Nicolet Groat, et al., "A Novel Gene Containing a Trinucleotide Repeat That Is Expanded and Unstable on Huntington's Disease Chromosomes," *Cell*, Vol. 72, No. 6, 1993.

[11] Online Mendelian Inheritance in Man, "OMIM Gene Map Statistics," webpage, May 4, 2020.

[12] Maria Jackson, Leah Marks, Gerhard H. W. May, and Joanna B. Wilson, "The Genetic Basis of Disease," *Essays in Biochemistry*, Vol. 62, No. 5, 2018.

wide association studies (GWAS), for finding the genetic basis of complex traits influenced by multiple genes.[13]

GWAS determine which genetic markers in a set occur more frequently among individuals with a phenotype, such as depression, than they do among individuals who do not exhibit the phenotype.[14] GWAS will commonly examine about one million single nucleotide polymorphism (SNP) genomic markers.[15] Although markers identify genomic locations important to a trait, other nearby genetic changes are often directly responsible for the contribution of the region to the trait. Thus, additional data and experiments are needed to find the causative genetic changes. While ability to attribute functions (e.g., regulatory regions, protein coding regions, etc.) to specific deoxyribonucleic acid (DNA) regions is improving, additional data and analysis are needed to fully understand the complexity of the human genome.[16] Also, as discussed, additional work is needed to understand the nongenetic contributions to traits and how nongenetic and genetic factors interact.

For example, a 2019 GWAS on insomnia used genetic data on 386,533 individuals from the United Kingdom Biobank and genetic data on 944,477 individuals from the 23andMe company to identify an association with

---

[13] Shiro Ikegawa, "A Short History of the Genome-Wide Association Study: Where We Were and Where We Are Going," *Genomics & Informatics*, Vol. 10, No. 4, 2012.

[14] William S. Bush and Jason H. Moore, "Chapter 11: Genome-Wide Association Studies," *PLOS Computational Biology*, Vol. 8, No. 12, 2012.

[15] An *SNP* is a single base within the human genome sequence that is highly variable within the population. For example, in one SNP location, 60 percent of the population might have an adenine (A) base while 40 percent of the population has a guanine (G) base. In another SNP location, 99 percent of the population might have a cytosine (C) base while 1 percent has a thymine (T) base. GWAS will commonly identify multiple (e.g., 25) markers with a statistically significant association with the trait of interest. The presence of a specific base in a statistically significant SNP location might, however, explain only several percent of the variation of the phenotype within the population. For example, if 10 percent of the overall population has a phenotype, 13 percent of the people with a statistically significant SNP (e.g., an A instead of a T in a particular location) might have the phenotype.

[16] Lee Siggens and Karl Ekwall, "Epigenetics, Chromatin and Genome Organization: Recent Advances from the ENCODE Project," *Journal of Internal Medicine*, Vol. 276, No. 3, 2014.

202 genomic regions, suggesting the involvement of more than 900 genes.[17] While the study serves as a valuable resource for future work on insomnia, the results are inadequate for identifying a set of genetic changes that would cure insomnia in an individual—even if the technology existed to make those changes (see next section) and addressing only the genetic contributions—rather than the environmental ones—was sufficient to make a positive impact.[18]

Limitations notwithstanding, multiple GWAS have examined topics of interest to improving human performance. For example, multiple studies have examined the genetic basis of hypoxia (oxygen deficiency) tolerance by examining the genetics of the Bajau people, an Indonesian hunter-gatherer population that collects its food through free diving,[19] and genetic adaptations in the Andean, Ethiopian, and Tibetan populations that live at high altitudes where oxygen is sparse.[20] Other studies considered a vari-

---

[17] Philip R. Jansen, Kyoko Watanabe, Sven Stringer, Nathan Skene, Julien Bryois, Anke R Hammerschlag, Christiaan A de Leeuw, Jeroen S Benjamins, Ana B Muñoz-Manchado, Mats Nagel, et al., "Genome-Wide Analysis of Insomnia in 1,331,010 Individuals Identifies New Risk Loci and Functional Pathways," *Nature Genetics*, Vol. 51, No. 3, 2019. Note that the large genetic data sets used—which can be reused to study a variety of phenotypes—enabled the identification of more genomic regions and involved genes compared with earlier GWAS.

[18] National Human Genome Research Institute and European Molecular Biology Laboratory—European Bioinformatics Institute, *GWAS Catalog*, database, undated. The 138,312 genetic associations cataloged in the *GWAS Catalog* as of May 3, 2019 (collected from 3,989 studies) also offer similar promise, potential, and challenges.

[19] Melissa A. Ilardo, Ida Moltke, Thorfinn S. Korneliussen, Jade Cheng, Aaron J. Stern, Fernando Racimo, Peter de Barros Damgaard, Martin Sikora, Andaine Seguin-Orlando, Simon Rasmussen, et al., "Physiological and Genetic Adaptations to Diving in Sea Nomads," *Cell*, Vol. 173, No. 3, 2018.

[20] Melissa Ilardo and Rasmus Nielsen, "Human Adaptation to Extreme Environmental Conditions," *Current Opinion in Genetics & Development*, Vol. 53, December 2018.

ety of phenotypes, such as hand grip strength,[21] intelligence,[22] information-processing speed,[23] radiation injury susceptibility,[24] athletic endurance,[25] posttraumatic stress disorder susceptibility,[26] and adaptation to arctic climates.[27] Researchers have also reviewed the growing literature and conducted non-GWAS analyses to identify genetic changes that might be advantageous for specific activities, such as sprinting,[28] or for thriving in

---

[21] Sara M. Willems, Daniel J. Wright, Felix R. Day, Katerina Trajanoska, Peter K. Joshi, John A. Morris, Amy M. Matteini, Fleur C. Garton, Niels Grarup, Nikolay Oskolkov, et al., "Large-Scale GWAS Identifies Multiple Loci for Hand Grip Strength Providing Biological Insights into Muscular Fitness," *Nature Communications*, Vol. 8, No. 16015, 2017.

[22] Suzanne Sniekers, Sven Stringer, Kyoko Watanabe, Philip R. Jansen, Jonathan R. I. Coleman, Eva Krapohl, Erdogan Taskesen, Anke R. Hammerschlag, Aysu Okbay, Delilah Zabaneh, et al., "Genome-Wide Association Meta-Analysis of 78,308 Individuals Identifies New Loci and Genes Influencing Human Intelligence," *Nature Genetics*, Vol. 49, No. 7, 2017.

[23] Carla A. Ibrahim-Verbaas, J. Bressler, Stéphanie Debette, Maaike Schuur, A. V. Smith, J. C. Bis, Gail Davies, Stella Trompet, J. A. Smith, C. Wolf, et al., "GWAS for Executive Function and Processing Speed Suggests Involvement of the CADM2 Gene," *Molecular Psychiatry*, Vol. 21, No. 2, 2016.

[24] Tong-Min Wang, Guo-Ping Shen, Ming-Yuan Chen, Jiangbo Zhang, Y. Sun, Jing He, Wen-Qiong Xue, Xi-Zhao Li, Shao-Yi Huang, Xiao-Hui Zheng, et al., "Genome-Wide Association Study of Susceptibility Loci for Radiation-Induced Brain Injury," *Journal of the National Cancer Institute*, Vol. 111, No. 6, 2018.

[25] Tuomo Rankinen, Noriyuki Fuku, Bernd Wolfarth, Guan Wang, Mark A. Sarzynski, Dmitry Alexeev, Ildus I. Ahmetov, Marcel R. Boulay, Pawel Cieszczyk, Nir Eynon, et al., "No Evidence of a Common DNA Variant Profile Specific to World Class Endurance Athletes," *PLoS One*, Vol. 11, No. 1, 2016.

[26] Sunayana B. Banerjee, Filomene G. Morrison, and Kerry J. Ressler, "Genetic Approaches for the Study of PTSD: Advances and Challenges," *Neuroscience Letters*, Vol. 649, 2017; Marilyn C. Cornelis, Nicole R. Nugent, Ananda B. Amstadter, and Karestan C. Koenen, "Genetics of Post-Traumatic Stress Disorder: Review and Recommendations for Genome-Wide Association Studies," *Current Psychiatry Reports*, Vol. 12, No. 4, 2010.

[27] Ilardo and Nielsen, 2018.

[28] Nir Eynon, Erik D. Hanson, Alejandro Lucia, Peter J. Houweling, Fleur C. Garton, Kathryn N. North, and David John Bishop, "Genes for Elite Power and Sprint Performance: ACTN3 Leads the Way," *Sports Medicine*, Vol. 43, No. 9, 2013.

situations in which humans have little experience, such as space flight.[29] In the future, the use of systems genetics techniques, which combine GWAS data with other large-scale data sets (e.g., gene expression, proteomics, metabolomics), will likely increase.[30]

Beyond looking at natural diversity that is present in the human population, researchers are also considering how genes from other organisms could provide capabilities beyond those for which humans have evolved. For example, Harvard Medical School's Consortium for Space Genetics is considering what genetic modifications might be needed to allow human populations to live in locations other than Earth.[31] Suggestions include adding genes from *Deinococcus radiodurans*, a bacterium that can survive in high levels of radiation, and adding genes from a variety of organisms to enable humans to synthesize all 20 amino acids (humans normally synthesize only 11 and extract the remaining nine from food).[32] Only a fraction of the possible protein folds have emerged through evolution, and techniques for modifying existing proteins and designing entirely new proteins are under way.[33] Identifying or designing relevant genes is just the first step; the genes would still need to be embedded within the existing human genetic circuitry, which might require additional genetic modifications. Scientists still do not know basic information about the function of approximately

---

[29] Michael A. Schmidt and Thomas J. Goodwin, "Personalized Medicine in Human Space Flight: Using Omics Based Analyses to Develop Individualized Countermeasures That Enhance Astronaut Safety and Performance," *Metabolomics*, Vol. 9, No. 6, 2013.

[30] Qingyang Huang, "Genetic Study of Complex Diseases in the Post-GWAS Era," *Journal of Genetics and Genomics*, Vol. 42, No. 3, 2015. Considering genetic changes in the context of human biological circuits should help researchers to understand why the identified genetic variants result in the observed phenotypes and to predict the impact of genetic changes.

[31] Harvard Medical School Consortium for Space Genetics, "About Us," webpage, undated.

[32] Jason Pontin, "The Genetics (and Ethics) of Making Humans Fit for Mars," *Wired*, July 7, 2018.

[33] Derek N. Woolfson, Gail J. Bartlett, Antony J. Burton, Jack W. Heal, Ai Niitsu, Andrew R. Thomson, and Chris W. Wood, "De Novo Protein Design: How Do We Expand into the Universe of Possible Protein Structures?" *Current Opinion in Structural Biology*, Vol. 33, August 2015.

20 percent of human genes, much less understand the myriad ways in which the products encoded by the human genomes interact, so many HPE objectives will remain infeasible for years to come.[34]

Types of Genetic Modifications and the Associated Clinical Techniques

Coincident with the growing understanding of the human genetic code, researchers have been developing techniques for making genetic modifications to humans. As the technology has matured, genetic modification techniques have been recognized for their potential not only to treat diseases but also to enhance humans; this has led some experts to grapple with associated ethical issues.[35] However, the vast majority of this attention has been focused on therapeutic applications, so this section of the report draws heavily on gene therapy literature. It should be noted that the focus of this report is on the technological aspects of what *might* be done rather than the ethical questions of what *should* be done. Other resources, such as the National Academies of Sciences, Engineering, and Medicine report titled *Human Genome Editing: Science, Ethics, and Governance* and the references contained therein, provide an overview of the ethical landscape.[36]

Early gene therapy clinical trials in the 1990s resulted in fatalities and serious side effects, but understanding of the foundational science has improved since then.[37] In 2003, Chinese regulators were the first to approve a gene therapy for use in humans—a treatment for a type of head and neck

---

[34] Valerie Wood, Antonia Lock, Midori Harris, Kim Rutherford, Jürg Bähler, and Stephen G. Oliver, "Hidden in Plain Sight: What Remains to Be Discovered in the Eukaryotic Proteome?" *Open Biology*, Vol. 9, No. 2, 2019.

[35] President's Council on Bioethics, *Beyond Therapy: Biotechnology and the Pursuit of Happiness*, Washington, D.C., October 2003.

[36] National Academies of Sciences, 2017.

[37] Cynthia E. Dunbar, Katherine A. High, J. Keith Joung, Donald B. Kohn, Keiya Ozawa, and Michel Sadelain, "Gene Therapy Comes of Age," *Science*, Vol. 359, No. 6372, January 2018.

cancer.[38] The FDA's first gene therapy approval came in 2017: a treatment for some forms of acute lymphoblastic leukemia.[39] The majority of gene therapy trials have attempted to treat cancer; the second biggest focus has been diseases caused by defects in a single gene.[40] Figure 2.1 shows the subjects of gene therapy trials—including trials to treat cancer and disorders of the

FIGURE 2.1

Ongoing Worldwide Clinical Trials as of March 2019 on Gene Therapy, Gene-Modified Cell Therapy, Cell Therapy, or Tissue Engineering

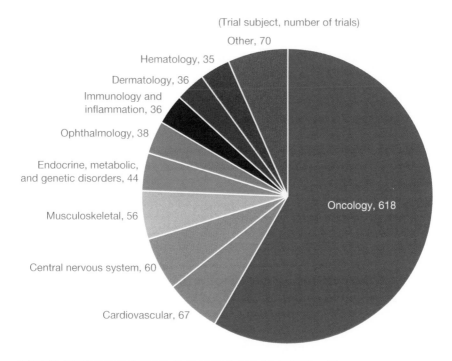

(Trial subject, number of trials)

Other, 70
Hematology, 35
Dermatology, 36
Immunology and inflammation, 36
Ophthalmology, 38
Endocrine, metabolic, and genetic disorders, 44
Musculoskeletal, 56
Central nervous system, 60
Cardiovascular, 67
Oncology, 618

SOURCE: RAND analysis of Alliance for Regenerative Medicine, 2019, p. 13.

---

[38] Sue Pearson, Hepeng Jia, and Keiko Kandachi, "China Approves First Gene Therapy," *Nature Biotechnology*, Vol. 22, No. 1, 2004.

[39] FDA, 2017.

[40] Samantha L. Ginn, Anais K. Amaya, Ian E. Alexander, Michael Edelstein, and Mohammad R. Abedi, "Gene Therapy Clinical Trials Worldwide to 2017: An Update," *Journal of Gene Medicine*, Vol. 20, No. 5, 2018.

cardiovascular system, the central nervous system, and the musculoskeletal system—being conducted in March 2019.[41]

Because of the paucity of work focused directly on improving human performance and the overall immaturity of the field, it is difficult to assess which researchers, organizations, or countries are in the lead for any particular trait. As noted, leveraging genetic manipulation to enhance human performance is a long-term possibility at best. The discussion provides context on areas of potential progress that might change the timeline unexpectedly.

Table 2.1 summarizes the main types of genetic modifications. Therapies that add a gene can treat diseases that arise when both copies of a gene in a patient (or a male patient's single copy of a gene on the X chromosome) are defective.[42] For diseases caused by the presence of a defective copy of a gene (rather than the absence of a functional copy of a gene), therapies that delete or edit the faulty gene are pursued.[43] Many variants are possible. For example, gene editing could be used to repair a nonfunctional gene. Although technologies for adding a gene are more mature than those for editing a gene or reducing the expression of a deleterious gene, both can result in unintended effects.[44] Gene therapy technologies are not sufficiently matured to make multiple, distinct changes simultaneously as would be necessary for many performance enhancement applications.

Another key attribute of a genetic change is its heritability (Table 2.2). Only heritable genetic changes can be passed to future generations. The first babies—twin girls—with edited genomes were born in 2018. A Chinese scientist, He Jiankui, had disabled both copies of the CCR5 gene, which confers HIV resistance, in one of the embryos.[45] The full effects of lack of CCR5 are unknown but likely include increased susceptibility to influenza, improved memory, and a shortened lifespan—all of which highlight the chal-

---

[41] Alliance for Regenerative Medicine, *Quarterly Regenerative Global Data Report: Q1 2019*, Washington, D.C., 2019, p. 13.

[42] Dunbar et al., 2018.

[43] Dunbar et al., 2018.

[44] Dunbar et al., 2018.

[45] David Cyranoski, "CRISPR-Baby Scientist Fails to Satisfy Critics," *Nature*, Vol. 564, No. 7734, 2018.

TABLE 2.1
Types of Genetic Modifications

Modification Type	Example Application	Limitations
Add a gene (e.g., insert gene permanently into recipient genome—usually at a random location; insert DNA element with gene into cell, but do not integrate into genome).	• Spinal muscular atrophy patients receive a replicative-defective virus containing the SMN1 gene (patients lack a functional copy) intravenously, extending lifespan and reducing symptoms.[a]	• Insertion technology limits size of gene to be added[b] • Insertion can damage or change the regulation of other genes, leading to serious side effects (e.g., cancer)[c] • Immune reactions in some patients to the vector used to introduce the gene[b]
Delete or modify a gene (i.e., change specific location of recipient genome in a specific way, including gene addition to specific genome location).	• In a mouse model, a defective version of the dystrophin gene (Duchenne muscular dystrophy patients lack a functional copy) was modified to create a partially functional version.[d]	• Introduction of changes to other genome locations[b] • Inaccurate editing of target location[b] • Challenges getting all components of the editing system into cell[b]
Add, delete, or modify multiple genes.	• Gene editing inactivated all 62 copies of a retrovirus in a porcine cell, which is a first step to making the cells suitable for use in human organ transplants.[e]	• Errors from each genetic alteration accumulate • Unless all locations to be modified are similar (as in example to left), prohibitively time-consuming and complex serial processing is required

[a] Jerry R. Mendell, Samiah Al-Zaidy, Richard Shell, W. Dave Arnold, Louise R. Rodino-Klapac, Thomas W. Prior, Linda Lowes, Lindsay Alfano, Katherine Berry, Kathleen Church, et al., "Single-Dose Gene-Replacement Therapy for Spinal Muscular Atrophy," *New England Journal of Medicine*, Vol. 377, No. 18, 2017.
[b] Dunbar et al., 2018.
[c] Matthew P. McCormack and Terence H. Rabbitts, "Activation of the T-Cell Oncogene LMO2 After Gene Therapy for X-Linked Severe Combined Immunodeficiency," *New England Journal of Medicine*, Vol. 350, No. 9, 2004.
[d] Christopher E. Nelson, Chady H. Hakim, David G. Ousterout, Pratiksha I. Thakore, Eirik A. Moreb, Ruth M. Castellanos Rivera, Sarina Madhavan, Xiufang Pan, F. Ann Ran, Winston X. Yan, et al., "In Vivo Genome Editing Improves Muscle Function in a Mouse Model of Duchenne Muscular Dystrophy," *Science*, Vol. 351, No. 6271, 2016.
[e] Luhan Yang, Marc Güell, Dong Niu, Haydi George, Emal Lesha, Dennis Grishin, John Aach, Ellen Shrock, Weihong Xu, Jürgen Poci, et al., "Genome-Wide Inactivation of Porcine Endogenous Retroviruses (PERVs)," *Science*, Vol. 350, No. 6264, 2015.

TABLE 2.2

Dimensions Characterizing Genetic Modifications

Attribute	Alternatives	Examples
Heritability	Yes—modification includes germline cells (i.e., sperm, egg, or embryo)	• The CCR5 gene was edited in a human embryo to confer HIV resistance, and then the embryo was implanted into a woman.[a]
	No—only somatic, not germline, cells modified	• Stem cells are removed from a sickle cell disease patient, modified by the addition of a normal gene encoding β-globin (the patient lacked a functional copy), and returned to the patient.[b]
Location of modification	In vivo (i.e., within organism)	• Cystic fibrosis patient inhales DNA encoding normal copy of CFTR gene (cystic fibrosis patients lack a functional copy) allowing the lung cells to temporarily make CFTR protein.[c]
	Ex vivo (e.g., in lab)	• T cells are removed from a leukemia patient, genetically modified to target leukemia cells, and returned to the patient.[d]

[a] Cyranoski, 2018.
[b] Ribeil et al., 2017
[c] Alton et al., 2015. CFTR = cystic fibrosis transmembrane conductance regulator.
[d] FDA, 2017.

lenges of identifying genetic changes that unambiguously confer improvements.[46] Heritable genetic modifications garner particular scrutiny because they can make enduring changes to the gene pool.

To enable the use of sophisticated laboratory procedures, the genetic modifications underlying some gene therapies are made *ex vivo*.[47] In *ex vivo* therapies, the target cells (e.g., immune system cells, stem cells, or embryos) are first removed from the body (or created via artificial fertilization or the creation of induced pluripotent stem cells outside the body). The desired modifications are then made in a laboratory setting. Finally, the modified cells are reintroduced to a human. *Ex vivo* therapies also enable a patient's own cells to be used in a treatment, removing the complications of finding a genetically matched donor.[48] Additionally, with an *ex vivo* approach, the modified cells can be tested for undesired genetic changes before they are placed in a human.

However, it is not practical to remove some types of cells from a patient to allow *ex vivo* manipulation. In those cases, *in vivo* gene therapy approaches are used to genetically modify cells while they are still in a patient's body. For example, *in vivo* approaches are used to provide gene therapies to the lungs of cystic fibrosis patients.[49] While *in vivo* approaches carry a risk of making heritable changes, unless germline cells are used, modifications made through *ex vivo* approaches are not heritable.[50]

Figure 2.2 shows the general steps in making a genetic modification, which are the same whether the goal is to improve performance or to cure a disease.[51] The first step is to determine the desired set of genetic modifications. As described earlier, GWAS can identify genetic markers that have statistically significant associations with the trait of interest. Follow-up work, which

---

[46] Sara Reardon, "Gene Edits to 'CRISPR Babies' Might Have Shortened Their Life Expectancy," *Nature*, Vol. 570, No. 7759, June 2019.

[47] Dunbar et al., 2018.

[48] Dunbar et al., 2018.

[49] Alton et al., 2015.

[50] Dunbar et al., 2018.

[51] A complete discussion of the wide variety of tools and approaches used in gene therapies and genetic modifications is beyond the scope of this chapter.

FIGURE 2.2

Steps in Making Genetic Modifications and Associated Foundational Technologies

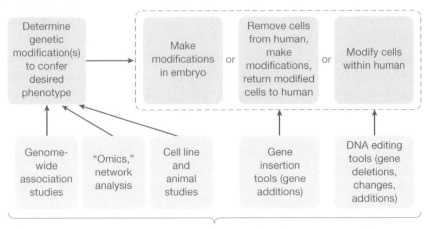

Foundational and enabling technologies

might focus on network analysis or cell line or animal studies, can then determine the specific DNA feature in the genomic region identified by the marker that contributes to the trait of interest.[52] The second step is then to make the desired genetic modifications, which are enabled by DNA editing tools.[53]

The near-term potential to use genetic modifications to confer a phenotype of interest on an individual depends on the number of DNA locations that contribute to the phenotype (see Figure 2.3). For phenotypes controlled

---

[52] Follow-up work is time-consuming and expensive, and it is typically pursued for only the DNA markers that explain the largest amounts of the trait's variability within the human population. Similarly, the fundamental understanding of human cellular networks needed to understand the secondary and tertiary impact of genetic changes (and identify compensatory genetic changes as needed) is still being developed.

[53] One option is to make the modifications within an embryo, which (barring off-target effects) will result in every cell of the resulting human receiving the modified DNA. Furthermore, that human will have the ability to pass the changes to his or her children. For other applications, such as immunotherapy, the cells to be modified are removed from a human, modified in a laboratory, and then returned to the human. In applications for which it is not practical to remove the cells to be modified from a human (e.g., lung tissue in a cystic fibrosis patient), the genetic modification can be done within the human and is typically targeted to cells in a specific tissue, such as the lungs.

by one DNA location (e.g., diseases caused by a single faulty gene), it is usually possible to find the causative DNA change (mutation). If the mutation acts primarily through its presence in a small number of cells that are accessible to genetic engineering (e.g., hemopoietic [blood] or immune system stem

FIGURE 2.3

Connection Between Number of DNA Locations Controlling a Trait and Current and Future Potential to Modify the Trait

	One or a few	Number of DNA locations that contribute to phenotype	Many
A) Example human traits controlled by indicated number of DNA locations	• Eye color • Huntington's disease • Sickle cell anemia • Cystic fibrosis	• Hemophilia • Immune cell targeting • Color blindness • Cheek dimples • Hair color	• Height • Strength • Intelligence • Depression • Asthma • Heart disease • Weight · Blood pressure · Cancer · Schizophrenia · Crohn's disease · Macular degeneration
B) State of the science for studying or modifying traits controlled by indicated number of DNA locations	• Find causative DNA change • *Functionally* correct some mutations in cells accessible to genetic engineering – Add correct version of gene while retaining nonfunctional copies – Disrupt defective copy of gene • Tolerate side effects of additional random DNA modifications		• Find multiple, not necessarily all, DNA regions associated with trait present in human population • Study identified DNA region to find single mutation responsible for portion of phenotype • *Functionally* correct single mutation causing large portion of phenotype (if any)
C) Possible future advances	• Reduce side effects of unintended DNA changes resulting from genetic modifications • Functionally correct mutations in more types of cells • Edit DNA to correct some mutations (e.g., convert defective gene to normal gene) in cells accessible to genetic engineering		• Design set of genetic changes to give nonhuman traits to humans, (e.g., ability to survive in space, infrared vision) • Understand all DNA locations that contribute to a trait and how changing a subset will affect a specific human • Develop tools easy and safe enough to make numerous genetic modifications

NOTES: **Row A** reflects example human traits controlled by small and large numbers of DNA locations. The example traits within each set are not ordered (i.e., those listed first in a box are not necessarily affected by fewer genes than those listed later in the box). In **Row B**, the bullets summarize current capabilities to determine the genetic basis of a trait and to change a human's DNA to affect the trait. The capabilities listed in the first column relate predominantly to traits controlled by one or a few DNA locations; the capabilities listed in the second column relate to traits affected by many DNA locations. The bullets in **Row C** summarize possible future advances in the field. The advances in the first column for traits controlled by no more than a few DNA locations will likely occur within the next five years; those related to traits controlled by multiple DNA locations (second column) will likely take longer.

23

cells that can be removed from the body, modified, and returned), then current techniques are often sufficient to functionally correct the mutation.[54] However, it is important to determine whether the benefits from correcting the mutation outweigh the risks of unintentionally creating other mutations as a side effect of the procedure. In the future, the number of unintended mutations should decrease and the number of tissues accessible for genetic engineering should increase. Within the next five to ten years, the ability to edit DNA precisely to convert a damaged sequence to the normal version should increase, although challenges will likely remain.[55]

Using genetic modifications to change traits controlled by multiple DNA locations is considerably more difficult.[56] When a defect in one gene is largely responsible for a disease in a patient (e.g., p53 in many cancers), even if numerous other mutations contribute, gene therapy can sometimes be used to target the responsible gene.

In contrast, traits of interest for HPE are typically influenced by numerous genetic locations, each with only a tiny impact. For example, a GWAS identified 18 genetic locations associated with intelligence, but those locations collectively explain only 4.8 percent of the variance in intelligence in the sample set (not all of which is heritable).[57] Thus, determining sets of genetic modifications to create such traits as high intelligence or high strength in specific individuals (or to create traits that are not present within the human population,

---

[54] For example, a working version of a gene could be added to a cell while retaining the two original, damaged copies of the gene. Alternatively, a mutated gene that produces a damaging product could be disrupted or deleted so that it does not produce any product, leaving a cell's second, normal copy of the gene as the only active version. Functional corrections allow a cell to produce a product that it previously could not or to stop producing a deleterious product. However, functional corrections do not necessarily change the underlying DNA sequence to match that of a typical human.

[55] Alliance for Regenerative Medicine, 2019.

[56] It is possible to use GWAS to find multiple, but not necessarily all, DNA markers associated with the presence or absence of a trait, such as macular degeneration, or the degree to which a person possesses a trait, such as height or intelligence. Follow-up work can then be used to determine the specific DNA locations that contribute to the trait. In practice, follow-up work, which is time-consuming and expensive, is typically pursued for only the DNA markers that explain the largest amounts of the trait's variability within the human population.

[57] Sniekers et al., 2017.

such as infrared vision) and then making the desired genetic modifications with a minimum of side effects is more than ten years into the future.

Survey of Foundational Activities Relevant to Improving Human Performance Through Genetic Modifications

To understand the extent to which different countries are acquiring foundational capabilities that might be applied to improving human performance through genetic modifications, relevant publications and clinical trials were culled from the open literature.

Figure 2.4 illustrates the number of articles on representative topics published January 1, 2016, through June 10, 2019, that are indexed by the Web of Science database.[58] The European Union led the world in the number of published gene therapy articles in that time frame.[59] The United States was a close second, and China was third. The United States led the world on articles about CRISPR. The European Union was second, and China was third. The European Union published almost twice as many articles on GWAS as the United States, the second most prolific publisher. China again was third. The quality and impact of published articles vary, but the number of articles published does provide a rough indicator of the amount of effort a country is devoting to a topic. However, Chinese articles in international journals are cited less frequently than the global average (citation frequency is associated with quality), so the gap in quality between China and the United States is likely larger than the difference in publication counts indicates.[60]

Table 2.3 shows the 20 most-common author affiliations of Web of Science articles on CRISPR. The Chinese Academy of Sciences and its associated uni-

---

[58] Clarivate, Web of Science database, undated.

[59] The United Kingdom was a member of the European Union at the time the analyses for this report were conducted, so European Union sets contain work from the United Kingdom.

[60] Futao Huang, "Quality Deficit Belies the Hype: Few Chinese Researchers Are Regarded as Global Leaders, as the Pressure for Rapid Output Prevails," *Nature*, Vol. 564, No. 7735, December 2018.

FIGURE 2.4

Country or Region of Author Affiliations of Papers Published in 2016 or Later on Gene Therapy, CRISPR, and Genome-Wide Association Studies

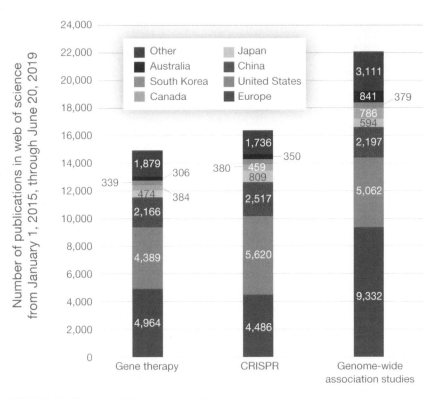

SOURCE: RAND analysis of Clarivate, undated.
NOTES: The analysis used the country associations in Web of Science's "Country/Region" field, which extracts the information from authors' addresses. A paper was counted with each country with which it was associated. The analysis disregarded papers without country information, which typically included 2–4 percent of the papers returned for a query. The "gene therapy" set came from a subject search for "gene therap*." The GWAS set came from a subject search for GWAS or "genome wide association stud*." The CRISPR set came from a subject search for "CRISPR*." Searches were performed on June 10, 2019.

versity were both among the five most prolific publishers. Japan's University of Tokyo was the only other non-U.S. institution among the top ten.

Multiple GWAS analyses have examined topics of interest related to improving human performance, and many other studies have built on the results of that work. Table 2.4 shows the five countries that have published

TABLE 2.3

Most-Frequent Author Affiliations for Web of Science Articles on CRISPR Published in 2016 or Later

Rank	Author Affiliation	Country
1	Chinese Academy of Sciences	China
2	Harvard Medical School	United States
3	Stanford University	United States
4	University of the Chinese Academy of Sciences	China
5	Massachusetts Institute of Technology (MIT)	United States
6	University of California, San Francisco	United States
7	University of California, Berkeley	United States
8	Harvard University	United States
9	University of California, San Diego	United States
10	University of Tokyo	Japan
11	Broad Institute of MIT and Harvard University	United States
12	University of Washington	United States
13	University of Pennsylvania	United States
14	Sun Yat-sen University	China
15	Seoul National University	South Korea
16	Chinese Academy of Agricultural Sciences	China
17	Shanghai Jiao Tong University	China
18	University of Cambridge	United Kingdom
19	Massachusetts General Hospital	United States
20	Peking University	China

SOURCE: Clarivate, undated.
NOTES: The set of papers came from a subject search for "CRISPR*" performed on June 16, 2019.
To increase the fraction of the papers that included original research, only publications of type
"Article" were included. (In contrast, Table 2.4 used data from all publications.)

the most papers indexed in the Web of Science that cite other papers on the genetic basis of hypoxia in humans, suggesting that they are engaged in

TABLE 2.4

Countries and Associated Organizations Publishing the Most Papers Indexed in Web of Science That Cite Other Papers on the Genetic Basis of Hypoxia in Humans

Rank–Country	Organizations in Country Publishing the Most Papers			
	1st	2nd	3rd	4th
1. United States	University of California System	Harvard University	National Institutes of Health	University of Kentucky
2. China	Jiangxi Agricultural University	Chinese Academy of Sciences	Shanghai Jiao Tong University	Army Medical University
3. England	University of Oxford	University of Cambridge	University of London	Imperial College of London
4. Denmark	University of Copenhagen	Aarhus University	Novo Nordisk	Novo Nordisk Foundation
5. Canada	University of Montreal	University of Toronto	University of British Columbia	University of Sherbrooke

SOURCE: RAND analysis of Clarivate, undated.

NOTE: The data consisted of the 626 Web of Science papers that cite a paper returned by a Web of Science subject search for (genome wide association stud or GWAS) and (hypoxia or altitude or "free diving") and (human* or people)). Searches were performed on July 12, 2019. The organizations in the table published 38 percent of the papers in the set. The search included the whole data range in the database, 1980 to the present. The analysis used the country associations in Web of Science's "Country/Region" field, which extracts the information from the records' addresses. A paper was counted with each country with which it was associated. The analysis disregarded papers without country information, which typically included 2–4 percent of the papers returned for a query. The analysis used Web of Science "Organizations-Enhanced" feature, which groups common variants of an organization's name, to determine the organizations that published the most papers for each country.

follow-up work. The table also shows the four organizations within each country that cited the most papers on the topic.

Like hypoxia, muscle strength is another topic of interest in the field of improving human performance. Table 2.5 shows the five countries that have cited the most papers indexed in the Web of Science that discuss the genetic basis of muscle strength in humans, suggesting that they are engaged in follow-up work. The table also shows the four organizations within each country that cited the most papers on the topic. Some of the institutions citing papers on work with humans are likely studying animals (e.g., Chinese Academy of Agricultural Sciences, Chinese Institute of Animal Science).

Figure 2.5 shows the number of clinical trials on selected topics that started in the three-year period between June 1, 2016, and May 31, 2019, and were registered with the U.S. National Library of Medicine.[61] More gene therapy and stem cell clinical trials occurred in the United States than any-where else in the world; Europe and China are second and third, respectively (Figure 2.5A and 2.5B). When only Phase 3 trials are considered (the final phase prior to regulatory approval), the United States, Europe, and China remain first, second, and third, respectively, but the Phase 3 trial dif-ferences among the United States, China, and Europe are less than when all clinical trials are considered (Figure 2.5D).

Figure 2.5C focuses on clinical trials that use CRISPR, which is widely considered to be the next step in the evolution of therapeutic gene editing.[62] Although counts are low, China started twice as many CRISPR trials in the three-year analysis period as the United States. Regulatory differences between the United States and China likely contribute to China's ability to fast-track clinical trials using this rapidly developing technology.[63]

Table 2.6 shows the sponsors and collaborators for gene therapy clinical trials that started in the three-year period beginning June 1, 2016 (compa-rable with the set shown in Figure 2.5A) and were registered with the U.S. National Library of Medicine. The U.S. National Cancer Institute was the

---

[61] U.S. National Library of Medicine, ClinicalTrials.gov, database, undated.

[62] Dennis Normile, "China Sprints Ahead in CRISPR Therapy Race," *Science*, Vol. 358, No. 6359, 2017.

[63] Normile, 2017.

TABLE 2.5

Countries and Associated Organizations Publishing the Most Papers Indexed in Web of Science That Cite Papers on the Genetic Basis of Muscle Strength in Humans

Rank-Country	Organizations in Country Publishing the Most Papers			
	1st	2nd	3rd	4th
1. United States	Harvard University	University of California System	VA Boston Healthcare System	National Institutes of Health
2. England	University of London	Newcastle University UK (United Kingdom)	Kings College London	University College London
3. China	Chinese Academy of Agricultural Sciences	Chinese Academy of Sciences	Institute of Animal Science	China Agricultural University
4. Canada	McGill University	Laval University	McMaster University	University of Alberta
5. Italy	Catholic University of the Sacred Heart	Consiglio Nazionale Delle Ricerche Cnr	IRCCS Policlinico Gemelli	Sapienza University Rome

SOURCE: RAND analysis of Clarivate, undated.

NOTES: The data consisted of 2,157 Web of Science papers that cite a paper returned by a Web of Science subject search for ((("genome wide association stud*" or GWAS or genomic*) and ("musc* strength" or "grip strength" or "muscle mass") and (human* or people)). Searches were performed on July 12, 2019.The organizations in the table published 37 percent of the papers in the set. The search included the whole data range in the database, 1980 to the present. The analysis used the country associations in Web of Science's "Country/Region" field, which extracts the information from the records' addresses. A paper was counted with each country with which it was associated. The analysis disregarded papers without country information, which typically included 2–4 percent of the papers returned for a query. The analysis used Web of Science "Organizations-Enhanced" feature, which groups common variants of an organization's name, to determine the organizations that published the most papers for each country.

FIGURE 2.5

Locations of Clinical Trials Starting Between June 1, 2016, and May 31, 2019

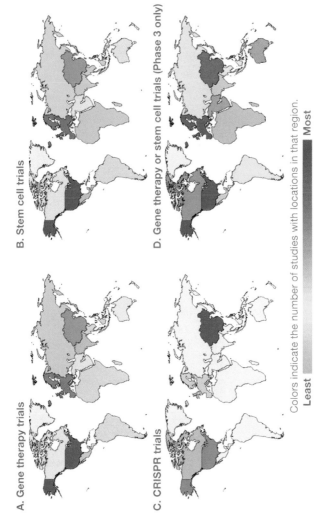

A. Gene therapy trials

B. Stem cell trials

C. CRISPR trials

D. Gene therapy or stem cell trials (Phase 3 only)

Colors indicate the number of studies with locations in that region.

Least   Most

SOURCE: U.S. National Library of Medicine, undated.

NOTE: The maps show the results of searching the "other terms" field of the database for (A) "gene therapy," (B) "stem cell," (C) "CRISPR," and (D) "gene therapy OR stem cell." Results shown in (D) were further filtered to include only Phase 3 trials. Searches were performed on June 9, 2019.

TABLE 2.6

Most-Frequent Sponsors or Collaborators of Gene Therapy Trials Starting Between June 1, 2016, and May 31, 2019

Trial Sponsor or Collaborator	Number of Trials	Country
National Cancer Institute	117	United States
AstraZeneca	20	United Kingdom
Baylor College of Medicine	17	United States
National Institutes of Health Clinical Center	17	United States
The Methodist Hospital System	14	United States
University of California, San Francisco	13	United States
Center for Cell and Gene Therapy, Baylor College of Medicine	13	United States
Assiut University	12	Egypt
M. D. Anderson Cancer Center	12	United States
Texas Children's Hospital	12	United States
Shenzhen Geno-Immune Medical Institute	11	China
Assistance Publique–Hôpitaux de Paris	11	France
Sun Yat-sen University	9	China
National Heart, Lung, and Blood Institute	8	United States
Jonsson Comprehensive Cancer Center	8	United States
Hospices Civils de Lyon	8	France
Duke University	7	United States
Fred Hutchinson Cancer Research Center	7	United States
University Health Network, Toronto	7	Canada
Dana-Farber Cancer Institute	7	United States

SOURCE: RAND analysis of data from U.S. National Library of Medicine, undated.

NOTES: The table displays results of searching the "other terms" field of the database for "gene therapy" on June 15, 2019, and tallying entries in the "Sponsor/Collaborators" field. Many trials listed multiple sponsors.

most common sponsor. The United Kingdom's AstraZeneca and Egypt's Assiut University were the only organizations outside the United States in the top ten.

In addition to the U.S. National Library of Medicine's clinical trials database, many countries have their own clinical trials databases. For example, the Chinese Clinical Trial Registry contains information about past, ongoing, and planned Chinese clinical trials.[64] As of July 12, 2019, the registry contained 24,289 trials (on all topics). The National Institutes of Health's Clinical Trials database,[65] which has entries for more than 300,000 trials, only contains 48 Chinese clinical trials database registrations.[66] Additional work would be needed to determine whether the National Institute of Health database contains additional trials in common with the Chinese registry that omitted their Chinese registry identifier.

Both the Chinese- and English-language search tools on the Chinese Clinical Trial Registry website allow users to search for trials using a set of criteria featuring title, disease, and sponsor. The search tools do have various limitations, however. Among them are

- no support for wildcards
- no support for Boolean expressions
- no support for phrases
- no ability to export a set of search results to a spreadsheet.

Table 2.7 shows some of the information available for a sample of gene therapy studies. Table 2.8 lists the main fields available in the database.

---

[64] Chinese Clinical Trial Registry, database, undated.

[65] U.S. National Library of Medicine, undated.

[66] U.S. National Library of Medicine, undated. A July 12, 2019, search of the database for "ChiCTR" (which is the common stem that the Chinese Clinical Trial Registry uses for its registration numbers) returned 48 results.

TABLE 2.7

Selected Fields for a Sample of Gene Therapy–Related Trials from the Chinese Clinical Trial Registry

Public Title	Scientific Title	Sources of Funding	Primary Sponsor	Target Disease	Objectives of Study
Gendicine intra-tumoral injection combined with radiotherapy for advanced cervical carcinoma	Recombinant adenoviral-p53 intra-tumoral injection combined with radiotherapy for advanced cervical carcinoma	Ministry of Science and Technique of China (973 project)	Beijing Cancer Hospital	Advanced cervical carcinoma	To evaluate the efficacy and adverse events of recombinant adenoviral-p53 intra-tumoral injection combined with radiotherapy for advanced cervical carcinoma
Clinical study for autologous RCAT cells plus anti-PD-1 antibody in patients with advanced renal cell carcinoma following failure of targeted therapy	Clinical study for combined immunotherapy in advanced renal cell carcinoma after failure of targeted therapy	Self-raising	He'nan Cancer Hospital	Renal carcinoma	Clinical efficacy and safety of RetroNectin and CD3 antibody activated cell plus anti-PD-1 antibody in patients with advanced renal cell carcinoma after failure of targeted therapy
Clinical study of recombinant human adenovirus type 5 (H101) combined with PD-1 antibody in the treatment of advanced solid tumors	Clinical study of recombinant human adenovirus type 5 (H101) combined with PD-1 antibody in the treatment of advanced solid tumors	Shanghai Sunway Biotech Co., Ltd	Second Hospital of Tianjin Medical University	Advanced solid tumors	To observe the efficacy and safety of recombinant human adenovirus type 5 (H101) combined with PD-1 antibody in the treatment of refractory tumors.

SOURCE: Chinese Clinical Trial Registry, undated.

34

TABLE 2.8

Overview of Main Fields in Chinese Clinical Trial Registry

Field	Subfield
Registration number	
Date of registration	Date of last refreshed on
Public title	
Scientific title	
Applicant	Telephone, fax, email, address, institution
Study leader	Telephone, fax, email, address
Ethics committee	
Primary sponsor	Address
Secondary sponsor	
Source(s) of funding	
Target disease	
Study type	
Study phase	
Objectives of study	
Study design	
Description for medicine or protocol of treatment in detail	Study design, inclusion/exclusion criteria, study execution time, interventions, countries of recruitment and research settings, randomization procedure, data collection and management

SOURCE: Chinese Clinical Trial Registry, undated.

Near-Term Trajectory for Genetic Modifications to Improve Human Performance and Implications for National Security

The bulk of the publications that examine national security implications of genetically modified humans focus on potential medium-term and long-term applications, such as creating designer warriors, and on the ethics of any application.[67] While any proposed activities should be considered in light of the Biological and Toxin Weapons Convention's prohibition of the use of "microbial or other biological agents, or toxins" for anything other than "prophylactic, protective or other peaceful purposes," additional deliberations are needed to define fully the convention's position on the topic of using genetic modifications to enhance humans.[68]

In the near future, the amount of foundational work to enable improvements in human performance through genetic modifications will likely continue increasing. The amount of data available to support GWAS efforts should continue growing as multiple countries—notably the United States, China, the United Kingdom, Saudi Arabia, the United Arab Emirates, and Turkey—have announced initiatives to sequence at least 100,000 genomes.[69] Nucleic acid testing has become a mechanism that China has leveraged in several countries, raising new questions about that nation's collection of data.[70] Similarly, the quality of the available gene editing tools should continue to improve, increasing editing efficiency and reducing off-target changes.[71] As more patients are treated with gene therapy, understanding of safety and efficacy will increase.

One area where work—at least, formally sanctioned work—will likely decrease (because of the negative global reaction to the birth of the first genetically modified children) is clinical work on modifying human embryos. In response to the 2018 birth of the first children from genetically modified embryos, prominent scientists have called for a worldwide

---

[67] For examples, see Allenby, 2018; and Greene and Master, 2018.

[68] Daniel Gerstein and James Giordano, "Rethinking the Biological and Toxin Weapons Convention?" *Health Security*, Vol. 15, No. 6, 2017; Convention on the Prohibition of the Development, Production and Stockpiling of Bacteriological (Biological) and Toxin Weapons and on Their Destruction, disarmament treaty, entered into force March 26, 1975.

[69] Alex Philippidis, "10 Countries in 100k Genome Club," *Clinical OMICs*, August 30, 2018.

[70] Sylvia Westall and Ivan Levingston, "Chinese Genetics Firm's Testing in Middle East Raises New U.S. Tensions," Bloomberg, May 20, 2020; Li Jianhua, "Coronavirus Pandemic Fuels China's Nucleic Acid Testing Industry," CGTN, February 22, 2021.

[71] National Academies of Sciences, Engineering, and Medicine, 2017, Appendix A.

moratorium on clinical human germline editing (i.e., work that results in the transfer of an edited embryo to a person's uterus) to give the world time to consider an appropriate governance framework.[72] Since then, the U.S. National Academy of Medicine, the U.S. National Academy of Sciences, and the United Kingdom's Royal Society convened the International Commission on the Clinical Use of Human Germline Genome Editing to develop a framework for assessing clinical applications of human germline editing.[73] By mid-2021, the World Health Organization developed recommendations for global standards for ethical gene editing.[74] However, gray-market work, of dubious quality, to provide genetically modified children for private clients might still occur.[75] Additionally, a Russian scientist has announced his intent to create additional genetically modified babies before governance regimes are modified to remove ambiguities.[76]

The current state of the science in gene therapy enables modifying single genes.[77] Performance enhancements to complex phenotypes (such as stamina or ability to work at high altitudes)—and particularly enhancements that minimize side effects—would require changes to multiple genes and an improved understanding of human biology.[78] For some phenotypes, "gene doping" treatments in specific adult tissues might be sufficient to temporally confer the phenotype on the recipient (analogous to how inhaling DNA encoding the CFTR protein can provide temporary treatment for cystic

---

[72] Eric S. Lander, Françoise Baylis, Feng Zhang, Emmanuelle Charpentier, Paul Berg, Catherine Bourgain, Bärbel Friedrich, J. Keith Joung, Jinsong Li, David Liu, et al., "Adopt a Moratorium on Heritable Genome Editing," *Nature*, Vol. 567, No. 7747, March 2019.

[73] National Academies of Sciences, Engineering, and Medicine, undated.

[74] Kate Goodwin, "China's CRISPR Babies Propel WHO to Issue Global Standards for Gene Editing," *BioSpace*, July 13, 2021.

[75] Antonio Regalado, "The DIY Designer Baby Project Funded with Bitcoin: Cryptocurrency, Biohacking, and the Fantastic Plan for Transgenic Humans," *MIT Technology Review*, February 1, 2019.

[76] Stepan Kravchenko, "Future of Genetically Modified Babies May Lie in Putin's Hands," Bloomberg, September 29, 2019.

[77] Dunbar et al., 2018.

[78] Huang, 2015.

fibrosis patients). For other phenotypes, the genetic modifications might need to be present in all cells from birth. Additional research is needed to identify the sets of genetic changes that might improve performance and to determine how to make them at no more than acceptable risk.

One application relevant to national security that should be possible in the near term is for militaries to perform GWAS analyses on the connections between warfighter genotype and performance on specific tasks, then (assuming that genetic variation explains a significant amount of variation in performance) assign tasks using the resulting model. For example, the U.S. military is exploring options, such as the use of genetic information, to better match warfighters to tasks.[79] A model can provide useful predictions without elucidating the underlying principles, so the additional understanding of human biology that is needed to posit beneficial genetic modifications is not needed for this application. As with all GWAS, however, it will be necessary to differentiate between genetic changes associated with a trait of interest (in this case, performance on a military task) and potentially confounding factors (e.g., genetic variation associated with race, performance variation resulting from variability in race-correlated opportunities for performance-enhancing preparatory experiences).

A related application also relevant to national security that scholars and ethicists have been considering since well before the current genomics revolution is the possibility of genetically modifying a bioagent to target a specific ethnic group or individual.[80] Although the availability of enabling data and methods is increasing, some steps are likely to remain difficult in the near term and it is unlikely that such a weapon would be able to sufficiently discriminate between target and nontarget populations.[81] To reduce the risk of legitimate research results being misapplied to such applications as ethnic weapons, many countries (including the United States) have adopted poli-

---

[79] Patrick Tucker, "Tomorrow Soldier: How the Military Is Altering the Limits of Human Performance," *Defense One*, July 12, 2017.

[80] For example, see British Medical Association, *Biotechnology, Weapons and Humanity*, London, Harwood Academic Publishers, 1999.

[81] Katherine Charlet, "The New Killer Pathogens: Countering the Coming Bioweapons Threat," *Foreign Affairs*, Vol. 97, No. 3, 2018, pp. 178–184.

cies to oversee dual-use research of concern.[82] Nonetheless, understanding of race and its relation to the host-pathogen relationship on a genetic and molecular level is improving:

Scientists discovered a genetic component to human susceptibility and resistance to many communicable diseases. For example, the contribution of the HBB (hemoglobin subunit beta) gene to malaria resistance was first reported in 1961, and subsequent studies have identified tens of additional genes that modulate either infection susceptibility or infection severity.[83] More recently, GWAS have identified genetic loci associated with a variety of common infectious diseases, such as chicken pox, strep throat, and scarlet fever.[84]

GWAS and genome sequencing studies have identified numerous genetic characteristics associated with race, although considerable population heterogeneity exists and the way that the bulk of the variability maps to phenotypes has not yet been explained.[85]

Techniques to map the *interactome* of a pathogen, the parts of a human cell that interact with the pathogen, are improving.[86] For example, interactome studies have identified thousands of human genes involved in influ-

---

[82] For example, see United States Government Policy for Institutional Oversight of Life Sciences Dual Use Research of Concern, September 24, 2014.

[83] Sandrine Marquet, "Overview of Human Genetic Susceptibility to Malaria: From Parasitemia Control to Severe Disease," *Infection, Genetics, and Evolution: Journal of Molecular Epidemiology and Evolutionary Genetics in Infectious Diseases*, Vol. 66, December 2018.

[84] Chao Tian, Bethann S. Hromatka, Amy K. Kiefer, Nicholas Eriksson, Suzanne M. Noble, Joyce Y. Tung, and David A. Hinds, "Genome-Wide Association and HLA Region Fine-Mapping Studies Identify Susceptibility Loci for Multiple Common Infections," *Nature Communications,* Vol. 8, No. 1, 2017.

[85] For example, see John Novembre, Toby Johnson, Katarzyna Bryc, Zoltán Kutalik, Adam R. Boyko, Adam Auton, Amit Indap, Karen S. King, Sven Bergmann, Matthew R. Nelson, et al., "Genes Mirror Geography Within Europe," *Nature*, Vol. 456, No. 7218, 2008; and Michael Yudell, Dorothy Roberts, Rob DeSalle, and Sarah Tishkoff, "Science and Society: Taking Race out of Human Genetics," *Science*, Vol. 351, No. 6273, 2016.

[86] Jamie Snider, Max Kotlyar, Punit Saraon, Zhong Yao, Igor Jurisica, and Igor Stagljar, "Fundamentals of Protein Interaction Network Mapping," *Molecular Systems Biology*, Vol. 11, No. 12, December 2015.

enza cellular entry and replication.[87] Work is also beginning to elucidate the genetic factors responsible for making a pathogen specific to a host and how a pathogen can evolve in a host during an infection.[88]

However, because the underlying biology is still not well understood, an actor seeking to create an ethnically targeted bioweapon would most likely need to test numerous pathogen variants in an experimental laboratory (not just a computational setting).[89] Although recent advances in high-throughput screening, biofoundries, and organoids could assist an actor with the testing, the likely need to conduct the experiments in biocontainment facilities would complicate and slow an attempt.

Additionally, it is not clear how much ethnic specificity a chosen pathogen could be modified to exhibit. That being the case, it is possible that none of the variants tested might have the sought-after properties. Thus, the prospect of a genetically enabled ethnic weapon is very slim in the near term and even the midterm.

---

[87] Tokiko Watanabe and Yoshihiro Kawaoka, "Influenza Virus-Host Interactomes as a Basis for Antiviral Drug Development," *Current Opinion in Virology*, Vol. 14, October 2015.

[88] For example, see Florian Douam, Jenna M. Gaska, Benjamin Y. Winer, Qiang Ding, Markus von Schaewen, and Alexander Ploss, "Genetic Dissection of the Host Tropism of Human-Tropic Pathogens," *Annual Review of Genetics*, Vol. 49, 2015; Pearson, Jia, and Kandachi, 2004; and Katherine S. Xue, Louise H. Moncla, Trevor Bedford, and Jesse D. Bloom, "Within-Host Evolution of Human Influenza Virus," *Trends in Microbiology*, Vol. 26, No. 9, 2018.

[89] Developing a pathogen tailored to harm a particular ethnic group would be very difficult. The possible experimental steps to construct a bioweapon and the potential for off-target effects are beyond the scope of this chapter.

CHAPTER THREE

Human Performance Enhancement and Artificial Intelligence

A High-Level Overview of Interactions Between Human Performance Enhancement and Artificial Intelligence

HPE fueled by AI (hereafter, HPE-AI) evokes striking images of the mental control of drone fleets and development of superhuman strength and intelligence. If successful, HPE-AI could substantially reduce the time required to process data and respond to situations. AI is being used to train BCIs, allowing users to manipulate neuroprosthetics, such as robotic limbs. The ways in which AI might improve the treatment of humans to enhance performance remains speculative. HPE-AI requires attention to human factors, customization, and training to get a reliable and useful pairing. These functions will add to the cost of fitting the system to each user and will add delay to the practical use of these systems.

Enhanced Performance Utilizing Brain Function

BCIs (see Figure 3.1) have been of research interest since the 1970s.[1] The Defense Advanced Research Projects Agency (DARPA)'s Augmented Cognition experiments were an effort to measure changes in human cognitive activ-

---

[1] Alexandre Gonfalonieri, "A Beginner's Guide to Brain-Computer Interface and Convolutional Neural Networks," *Towards Data Science,* November 25, 2018.

FIGURE 3.1

Diagram of Neuroimaging Methods

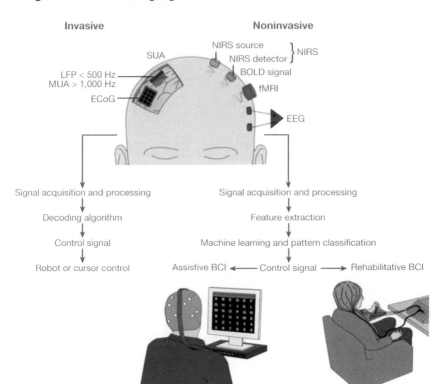

SOURCE: Reprinted from Ujwal Chaudhary, Niels Birbaumer, and Ander Ramos-Murguialday, "Brain–Computer Interfaces for Communication and Rehabilitation," *Nature Reviews Neurology*, Vol. 12, 2016, p. 515, with permission from Elsevier.

ity during execution of a task;[2] this information provides the signal parameters for the control signals in BCI-based movement of external objects. However, the rise of AI, machine-learning, deep neural nets, and hybrid algorithms has expanded data-processing for a variety of applications. The primary application researched for BCIs is neuroprosthetics (i.e., bypassing the nervous system for the purpose of controlling external apparatuses, such as mechani-

[2] Mark St. John, David A. Kobus, Jeffrey G. Morrison, and Dylan Schmorrow, "Overview of the DARPA Augmented Cognition Technical Integration Experiment," *International Journal of Human-Computer Interaction*, Vol. 17, No. 2, 2004.

cal limbs or cochlear implants). Humans can be trained to control all sorts of external devices, such as movement of a computer mouse,[3] robotic arms,[4] and drones.[5] A recent report indicates that BCIs coupled with recording of muscle activity in the face and neck allows synthesis of speech without vocalization by the person.[6] Nearly all the applications discussed in this report have been developed for therapeutic or restorative purposes, with the notable exceptions of DARPA's Augmented Cognition program and the recent Neuralink announcement (discussed later in this section).

Several mechanisms for noninvasive collection of brain activity require large, bulky equipment, specialized facilities, and stationary subjects to collect reasonable data—factors that militate against broad use. The techniques, which are limited (those listed as nonportable in Table 3.1), are fMRI, emission computed tomography (including PET), and MEG. The remaining noninvasive techniques are optical detection of blood flow and changes in blood oxygenation using "near infrared" (NIR) scanning or detection of patterns of electrical activity using EEG, which is the most commonly used technique for BCIs because of better spatial and temporal resolution.[7] Advances in EEG include smaller, portable alternatives that lack the ability to spatially resolve brain signals and instead rely on mathematical decon-

---

[3] Jonathan R. Wolpaw and Dennis J. McFarland, "Control of a Two-Dimensional Movement Signal by a Noninvasive Brain-Computer Interface in Humans," *Proceedings of the National Academy of Sciences*, Vol. 101, No. 51, 2004.

[4] Xiaogang Chen, Bing Zhao, Yijun Wang, Shengpu Xu, and Xiaorong Gao, "Control of a 7-DOF Robotic Arm System with an SSVEP-Based BCI," *International Journal of Neural Systems*, Vol. 28, No. 8, 2018.

[5] Karl LaFleur, Kaitlin Cassady, Alexander Doud, Kaleb Shades, Eitan Rogin, and Bin He, "Quadcopter Control in Three-Dimensional Space Using a Noninvasive Motor Imagery-Based Brain–Computer Interface," *Journal of Neural Engineering*, Vol. 10, No. 4, 2013.

[6] Gopala Krishna Anumanchipalli, Josh Chartier, and Edward F. Chang, "Speech Synthesis from Neural Decoding of Spoken Sentences," *Nature*, Vol. 568, No. 7753, 2019.

[7] Note that open source EEG data sets are available. They lack the context and range of variability that are necessary to effectively train a system, but they might be useful in developing training protocols.

TABLE 3.1
Neuroimaging Methods to Collect Brain Activity

Neuroimaging Method	Activity Measured	Risk	Spatial Resolution	Temporal Resolution	Portability
EEG	Electrical	Noninvasive	~10 mm	~0.001 second	Portable
Electrocorticography (ECoG)	Electrical	Semiinvasive	~1 mm	~0.003 second	Portable
MEG	Magnetic	Noninvasive	~5 mm	~0.05 second	Nonportable
PET	Metabolic	Noninvasive	~1 mm	~0.2 second	Nonportable
Single photon emission computed tomography (SPECT or SPET)	Metabolic	Noninvasive	~1 cm	~10 seconds–30 minutes	Nonportable
fMRI	Metabolic	Noninvasive	~1 mm	~1 second	Nonportable
Optical imaging (functional near infrared [fNIR])	Metabolic	Noninvasive	~2 cm	~1 second	Portable

SOURCE: Adapted from Rabie A. Ramadan and Athanasios V. Vasilakos, "Brain Computer Interface: Control Signals Review," *Neurocomputing*, Vol. 223, 2017, p. 30, with permission from Elsevier.

volution.[8] The other advances include more-transparent electrodes so that concurrent brain imaging can be used.[9]

There are more than a dozen different commercially available electrodes sets for recording EEG.[10] The devices differ in the number of channels that they record on, the resolution of the signal that is captured (i.e., the analog-to-digital converter), and the rate of sampling from the electrodes. The number and placement of channels affect spatial resolution. Analog-to-digital conversion limits the size of the signal that can be detected; the signals are binned according to size, and smaller signals are grouped with background electrical activity. The sampling rate is the temporal resolution captured. The commercial systems are frequently marketed for use in neurobiofeedback, a system used to learn to control brain electrical activity through the visualization (or auditory signal production) related to brain activity. Although neurobiofeedback has been used clinically in the treatment of attention deficit hyperactivity disorder, anxiety, and depression, the literature on its benefits is inconclusive;[11] biofeedback for pain management has a long history.[12] Generally, only a select few of the biofeedback systems designed to improve performance have been tested in healthy populations.[13]

One contemporary example is Muse, which is produced by a Canadian company established in 2007, primarily marketed for use with biofeedback

---

[8] Robert T. Mueller, "New EEG Technology Makes for Better Brain Reading," *Psychology Today,* September 18, 2014.

[9] Yi Qiang, Pietro Artoni, Kyung Jin Seo, Stanislav Culaclii, Victoria Hogan, Xuanyi Zhao, Yiding Zhong, Xun Han, Po-Min Wang, Yi-Kai Lo, et al., "Transparent Arrays of Bilayer-Nanomesh Microelectrodes for Simultaneous Electrophysiology and Two-Photon Imaging in the Brain," *Science Advances,* Vol. 4, No. 9, September 2018.

[10] See Ramadan and Vasilakos, 2017, Table 6.

[11] Hengameh Marzbani, Hamid Reza Marateb, and Marjan Mansourian, "Neurofeedback: A Comprehensive Review on System Design, Methodology and Clinical Applications," *Basic and Clinical Neuroscience,* Vol. 7, No. 2, 2016.

[12] Mayo Clinic, Biofeedback, webpage, undated; and Malik Kashif and Anterpreet Dua, "Biofeedback," StatPearls, December 20, 2019.

[13] Jonathan M. Peake, Graham Kerr, and John P. Sullivan, "A Critical Review of Consumer Wearables, Mobile Applications, and Equipment for Providing Biofeedback, Monitoring Stress, and Sleep in Physically Active Populations," *Frontiers in Physiology,* Vol. 9, June 2018.

systems,[14] and made in China.[15] Another example is Emotiv, Inc., which produces the Epoc and the Insight devices. Tan Le founded Emotiv Systems in 2003 in Australia and Emotiv, Inc., in 2011 in San Francisco. The U.S. company has facilities in Sydney, Australia, and in Hanoi and Ho Chi Minh City in Vietnam. NeuroSky is a privately held U.S. company whose products are manufactured in China.[16] OpenBCI develops biosensing hardware for researchers, makers, and hobbyists. It is unclear where the components are made for OpenBCI,[17] but the company has plans available for a 3D-printed EEG headset. OpenBCI was started with a 2013 Kickstarter campaign and had 17 repositories on GitHub in August 2021. There is significant use of the OpenBCI platform for research (565 results in GoogleScholar since 2020).

In summary, BCIs have appeal for complex, real-time, hands-free control of devices or robots. BCI devices and software tools are available commercially. A major limitation with BCI-controlled external devices is the variability in the training process. The effective signals that correlate with the intended motion of an external device are nonstationary.[18] This means that the signal can move to different parts of the surface of the brain. In addition, individuals vary in terms of where the signals are likely to be located, how the signals move, and the efficacy with which they can transfer information to the device.[19] Finally, the use of AI for tuning BCI applications is constrained by difficulties in developing labeled data for supervised training of the system and in validating to ensure the trained system behaves as expected.

---

[14] Muse, sales webpage, undated.

[15] UserManual.wiki, "Interaxon MU02 Bluetooth LE Device User Manual R1," webpage, undated.

[16] NeuroSky, "Enabling Technologies for Next-Generation mHealth Solutions," webpage, undated.

[17] The OpenBCI website cautions that shipments outside the United States might incur import fees, suggesting U.S. production. OpenBCI, "Shipping and Taxes," webpage, April 11, 2020.

[18] Haider Raza, Dheeraj Ratheeb, Shang-Ming Zhou, Hubert Cecotti, and Girijesh Prasad, "Covariate Shift Estimation Based Adaptive Ensemble Learning for Handling Non-Stationarity in Motor Imagery Related EEG-Based Brain-Computer Interface," *Neurocomputing*, Vol. 343, May 2019.

[19] Anumanchipalli, Chartier, and Chang, 2019; Wolpaw and McFarland, 2004.

A RAND project developed a national security game to explore the use of BCI in future combat scenarios and detailed the potential risks associated with its application.[20] This project found that the utility for BCI in combat is likely to increase as military applications of AI and robotics develop further. The research team concluded that the application of BCI would support ongoing U.S. Department of Defense technological initiatives, such as human-machine collaboration for improved decisionmaking, assisted-human operations, and advanced manned and unmanned combat teaming. The operational risks associated with the development and application of BCI in a combat setting are highlighted, and the ethical and legal challenges for the U.S. Department of Defense are noted. The data set used to train activation of the external device is based on a set of recorded brain activity. Those data can be collected through invasive or noninvasive means (see Figure 3.1 and Table 3.1). Invasive processes require surgical implantation of electrodes but provide a cleaner and more reliable signal of brain activity.[21] Invasive means to collect brain activity are typically limited to those instances for which the medical intervention is justified, often in patients who have suffered a stroke. However, Elon Musk's Neuralink company proposes implantation of micron-scale electrodes directly into the brain,[22] and the Center for Bioelectric Interfaces in Russia has focused on the development of electrode nets outside the skull but under the skin.[23] The measures increase the strength of signals captured from the brain and could improve the extent of the brain area that is recorded (better signal, more electrodes, more widely spaced). Russian researchers have focused on improving the signal from EEG and on detection of subjects' change in focus.[24] Under-

---

[20] Anika Binnendijk, Tim Marler, and Elizabeth M. Bartels, *Brain-Computer Interfaces: U.S. Military Applications and Implications, An Initial Assessment*, Santa Monica, Calif.: RAND Corporation, RR-2996-RC, 2020.

[21] Ramadan and Vasilakos, 2017.

[22] Elizabeth Lopatto, "Elon Musk Unveils Neuralink's Plans for Brain-Reading 'Threads' and a Robot to Insert Them," *The Verge,* July 16 2019.

[23] Centre for Bioelectric Interfaces, homepage, undated.

[24] Alexander E. Hramov, Vladimir A. Maksimenko, and Marina Hramova, "Brain-Computer Interface for Alertness Estimation and Improving," *Dynamics and Fluctuations in Biomedical Photonics XV Proceedings*, Vol. 10493, May 21, 2018; Alexander E.

standing fatigue or drops in focus for subjects using BCI is critical because of the continuous data collection from the EEG. For effective training and for real-time use of BCI, the EEG must reflect when the subject is focused on control of the computer system.

Chinese, Russian, and U.S. researchers and entrepreneurs have plotted different courses for the development of BCI. Here are some illustrations:

- In May 2019, China presented a BCI chip called "BrainTalker" at a conference in China.[25] The chip is designed to process electrical signals from the brain without a computer. The issue of the placement and efficiency of electrodes was not discussed.

- The Russian developers of Neurochat, a BCI kit based on EEG, are investigating whether the biofeedback from these systems improve rehabilitation of patients with motor or speech impairments.[26] Russian researchers at the Center for Bioelectric Interfaces argue that EEG is not effective for sustained control of external devices, but that implanted electrodes have only short-term functionality because of electrode fouling and high risks of complications.[27] Therefore, they have focused on development of subdural or epidural electrode nets for BCI.

- Elon Musk's Neuralink company has developed a robotic system for the implantation of thousands of micron-scale electrodes inside the brain for the stated purposes of using brain recordings for control of external devices and of electrically stimulating the brain.[28]

Hramov, Vladimir A. Maksimenko, Maxim D. Zhuravlev, and Alexander N. Pisarchik, "Immediate Effect of Neurofeedback in Passive BCI for Alertness Control," paper presented at the 7th International Winter Conference on Brain-Computer Interface (BCI), High 1 Resort, Korea, February 2019; Vladimir A. Maksimenko, Alexander E. Hramov, Vadim V. Grubov, Vladimir O. Nedaivozov, Vladimir V. Makarov, Alexander N. Pisarchik, "Nonlinear Effect of Biological Feedback on Brain Attentional State," *Nonlinear Dynamics*, Vol. 95, No. 3, 2019.

[25] "China Unveils Brain-Computer Interface Chip," Xinhua Net, May 18, 2019.

[26] "Brain-Controlled System Neurochat Begins to Be Batch-Produced in Russia," TASS Russian News Agency, April 23, 2019.

[27] Centre for Bioelectric Interfaces, undated.

[28] Lopatto, 2019.

- DARPA has developed a program to accelerate development of next-generation nonsurgical neurotechnology (N3), which focuses on non-invasive high-fidelity effective BCI with the performance of implanted electrodes.[29] The approaches vary, from using interference from electrical waves as stimulation to using a viral vector to incorporate and signal protein into neurons.

These examples document investment and exploration. It is premature to calibrate effectiveness or to predict the timing of practically usable capabilities.

AI to Process Individual Data and Diagnose Human Performance Deficits

The uses and benefits of AI to alter treatment of humans for enhanced performance largely remain areas of speculation. The pattern recognition and data-processing power of AI lends two potential advantages. First, the speed of data acquisition and processing would allow for real-time feedback, such as transcranial stimulation, to improve real-time human response and decisionmaking. BCI can be used to stimulate the central nervous system: the brain, spinal cord, and neurosensory retina. Devices with electrodes can be designed to deliver a neural command or to achieve general neural stimulation.[30] One review suggested that neural stimulation in association with training can result in improved motor learning, motion perception, or muscular strength or in reduced muscle fatigue.[31] There is interest in how machine-learning could inform and integrate neural stimulation protocols to improve the reported performance gains associated with neural stimu-

---

[29] Megan Scudellari, "DARPA Funds Ambitious Brain-Machine Interface Program," *IEEE Spectrum*, May 21, 2019.

[30] Gabriel A. Silva, "A New Frontier: The Convergence of Nanotechnology, Brain Machine Interfaces, and Artificial Intelligence," *Frontiers in Neuroscience*, Vol. 12, 2018.

[31] Lorenza S. Colzato, Michael A. Nitsche, and Armin Kibele, "Noninvasive Brain Stimulation and Neural Entrainment Enhance Athletic Performance—A Review," *Journal of Cognitive Enhancement*, Vol. 1, No. 1, 2017.

lation. Other research suggests that intersubject variability is high enough and intersubject reliability is low enough that the claims of cognitive and behavioral performance are not justified.[32]

Second, AI can be used to individualize treatment that is generally believed to affect performance positively, analogous to precision (personalized) medicine; personalization could be used with any performance enhancement modality.[33] Deep-learning and big data are having an impact across the medical field.[34] One anticipated outcome is individualized treatment fueled by patients' awareness of and access to their own medical data. AI could be used to characterize unique interventions that could also improve performance for individuals, given that this approach is being used to develop personalized diagnoses and treatments.[35] The use of AI could be combined with individual data to prescribe personalized performance management. Tailored microbiome manipulation is another area of interest to enhance performance and health.[36] A repository of classification tasks from 15 human microbiome data sets has been developed to support machine-learning for prediction of health outcomes from microbiome composition.[37] These data sets enable the use of AI to tailor individual microbiome supplements for enhanced performance.

China has made significant investment in precision medicine that has implications for performance management. Chinese-based BGI is among

---

[32] Jared C. Horvath, Olivia Carter, and Jason D. Forte, "Transcranial Direct Current Stimulation: Five Important Issues We Aren't Discussing (But Probably Should Be)," *Frontiers in Systems Neuroscience*, Vol. 8, No. 2, January 2014.

[33] Army Science Board, *Army Efforts to Enhance Soldier and Team Performance*, Washington, D.C., 2017.

[34] Eric J. Topol, "High-Performance Medicine: The Convergence of Human and Artificial Intelligence," *Nature Medicine*, Vol. 25, No. 1, 2019.

[35] Bertalan Mesko, "The Role of Artificial Intelligence in Precision Medicine," *Expert Review of Precision Medicine and Drug Development*, Vol. 2, No. 5, September 2017.

[36] Rachel N. Carmody and Aaron L. Baggish, "Working Out the Bugs: Microbial Modulation of Athletic Performance," *Nature Metabolism*, Vol. 1, 2019.

[37] Pajau Vangay, Benjamin M.Hillmann, and Dan Knights, "Microbiome Learning Repo (ML Repo): A Public Repository of Microbiome Regression and Classification Tasks," *GigaScience*, Vol. 8, No. 5, May 2018.

the world leaders in next-generation sequencing, storage of sequence data, and cloud computing.[38] BGI has offices in seven countries, including a U.S. computing facility in Silicon Valley. A sponsored supplement to *Science* magazine in 2018 focused on Chinese efforts in the areas of cancer immunology and precision medicine.[39] The precision medicine investment is related to the significant Chinese investment in biotechnology, including shares in several U.S. companies.[40] There is some evidence of Russian investment in precision medicine, from recent publications advocating its benefits to conferences sponsored by the Russian Federation.[41] The collection of vast amounts of personalized data enables a shift from looking for symptoms and tests to indicate a diagnosis to a system of data analysis for "prevention, personalization, and precision."[42]

---

[38] Benjamin Shobert, "Meet the Chinese Company That Wants to Be the Intel of Personalized Medicine," *Forbes*, January 18, 2017.

[39] Sean Sanders and Jackie Oberst, eds., *Sponsored Collection: Precision Medicine and Cancer Immunology in China*, supplement to *Science* magazine, Vol. 359, No. 6375, February 2, 2018.

[40] Shannon Ellis, "China's Fledgling Biotech Sector Fizzes into Life," *Nature Biotechnology*, Vol. 36, 2018a.

[41] Sergey Suchkov, H. Abe, E. N. Antonova, P. Barach, B. T. Velichkovskiy, M. M. Galagudza, D. A. Dworaczyk, D. Dimmock, V. M. Zemskov, I. E. Koltunov, et al., "Personalized Medicine as an Updated Model of National Health-Care System, Part 1: Strategic Aspects of Infrastructure," *Rossiyskiy vestnik perinatologii i pediatrii [Russian Bulletin of Perinatology and Pediatrics]*, Vol. 62, No. 3, 2017a; Sergey Suchkov, H. Abe, E. N. Antonova, P. Barach, B. T. Velichkovskiy, M. M. Galagudza, D. A. Dworaczyk, D. Dimmock, V. M. Zemskov, I. E. Koltunov, et al., "Personalized Medicine as an Updated Model of National Health-Care System, Part 2: Towards Public and Private Partnerships," *Rossiyskiy vestnik perinatologii i pediatrii [Russian Bulletin of Perinatology and Pediatrics]*, Vol. 62, No. 4, 2017b; World Health Organization, "Can Personalized Medicine Contribute to Prevention and Control of NCDs in the Russian Federation?" press release, May 14, 2018.

[42] Mesko, 2017.

Improving Individual Performance Through Human-AI Teaming

Much of the emphasis on the use of AI by the military has been on increasing the speed with which data are analyzed and acted on.[43] Speeding the decision cycle to the point at which an aided human can no longer keep up has been termed *hyperwar*. The pace of hyperwar requires utilization of intelligent decision support systems (IDSS), which have specialized functions (e.g., intelligent agents) that perform tasks related to decisionmaking. These functions include data mining and machine-learning, knowledge representation, intent recognition, and automated inference.[44] The IDSS have the advantage of being able to manage many more variables than humans, to identify potential cognitive biases, and to improve situational understanding.

Whether technology to support human-machine teaming can evolve fast enough and with sufficient effectiveness is an open question.[45] A recent discussion convened at RAND that focused on AI decision support systems and the potential impact on nuclear war cautioned against using AI as a trusted adviser because of the potential to exacerbate nuclear tensions.[46] Human-AI teams designed to function in high-stakes environments, such as military decisionmaking, need to be able to perform reliably in various circumstances and conditions. Humans working directly with an AI system need to understand the range of performance of that system. AI systems are designed to update with experience, including failures. A recent study on updates to AI systems determined that although those updates improve the

---

[43] Julian E. Barnes and Josh Chin, "The New Arms Race in AI," *Wall Street Journal*, March 2, 2018.

[44] Karel van den Bosch and Adelbert Bronkhorst, "Human-AI Cooperation to Benefit Military Decision Making," *Proceedings of the NATO IST-160 Specialist Meeting on Big Data and Artificial Intelligence for Military Decision Making*, Bordeaux, France, 2018.

[45] Human AI teams had been described as "centaurs" in the time of Garry Kasparov's and IBM's Deep Blue chess match. Nicky Case, "How to Become a Centaur," *Journal of Design and Science*, January 8, 2018.

[46] Edward Geist and Andrew J. Lohn, *How Might Artificial Intelligence Affect the Risk of Nuclear War?* Santa Monica, Calif.: RAND Corporation, PE-296-RC, 2018.

system's performance,[47] they also change the performance of the AI system, reducing the human users' trust of the output of the improved AI system. So, even though updates improved AI performance, the human-AI team performance was diminished (Figure 3.2). General principles for human-AI interaction have been discussed in the literature.[48] Generally applicable design guidelines for human-AI interaction have been proposed recently.[49] The 18 guidelines are in Table 3.2.

Careful consideration needs to be given to the human factors that ultimately determine the way that human-AI teams function in complex environments. *Human factors* is defined by the American National Standards Institute and the Association for the Advancement of Medical Instrumentation as "the application of knowledge about human capabilities (physical, sensory, emotional, and intellectual) and limitations to the design and development of tools, devices, systems, environments and organizations."[50] Human factors ultimately determine the range over which the AI is trusted to perform in a reliable way. This trusted range limits the extent to which the human-AI team outperforms conventional AI. Explainable AI (XAI) often requires that the algorithm being used include "explicit declarative knowledge" so that humans can understand the basis of the decisionmaking.[51] The degree of human understanding of the XAI influences not only

---

[47] Gagan Bansal, Besmira Nushi, Ece Kamar, Daniel S. Weld, Walter S. Lasecki, and Eric Horvitz, "Updates in Human-AI Teams: Understanding and Addressing the Performance/Compatibility Tradeoff," *Proceedings of the AAAI Conference on Artificial Intelligence*, Vol. 33, No. 1, July 17, 2019.

[48] Saleema Amershi, Dan Weld, Mihaela Vorvoreanu, Mihaela Vorvoreanu, Adam Fourney, Besmira Nushi, Penny Collisson, Jina Suh, Shamsi T. Iqbal, Paul N. Bennett, Kori Marie Inkpen, et al., "Guidelines for Human-AI Interaction," *Proceedings of the 2019 CHI Conference on Human Factors in Computing Systems*, Glasgow, Scotland, 2019.

[49] Amershi et al., 2019.

[50] American National Standards Institute and Association for the Advancement of Medical Instrumentation, *Human Factors Engineering—Design of Medical Devices*, New York, preview edition, 2013.

[51] Andreas Holzinger, Peter Kieseberg, Edgar Weippl, and A. Min Tjoa, "Current Advances, Trends and Challenges of Machine Learning and Knowledge Extraction: From Machine Learning to Explainable AI," paper presented at International Cross-Domain Conference, Hamburg, Germany, August 2018.

FIGURE 3.2

Schematized View of Human-AI Teams in the Presence of AI Updates

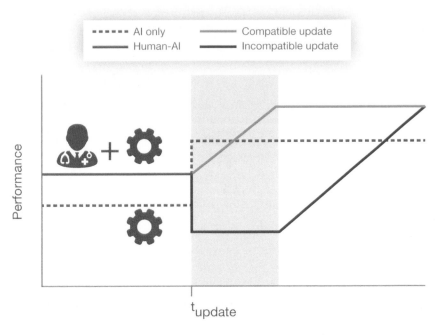

SOURCE: Bansal et al., 2019. Used with permission.

human understanding of the problem but also human trust in and use of the AI.[52] DARPA's XAI program was designed to foster algorithms that consider the context and environment in which they are operated.[53] Broad AI teaming has been predicted to increase the efficiency and speed of decisionmaking in war. Researchers and ethicists are considering the circumstances (if any) under which AI systems should be allowed to make lethal decisions and whether, with proper programming, AI systems might be able to behave

---

[52] Robert R. Hoffman, Gary Klein, and Shane T. Mueller, "Explaining Explanation for 'Explainable AI,'" *Proceedings of the Human Factors and Ergonomics Society Annual Meeting*, September 25, 2018.

[53] Matt Turek, "Explainable Artificial Intelligence (XAI)," DARPA program information, undated.

TABLE 3.2

Design Guidelines for Effective Human-AI Teaming

		All design guidelines	Example applications of guidelines
Initially	G1	**Make clear what the system can do.** Help the user understand what the AI system is capable of doing.	[Activity Trackers, Product #1] "Displays all the metrics that it tracks and explains how. Metrics include movement metrics such as steps, distance traveled, length of time exercised, and all-day calorie burn, for a day."
	G2	**Make clear how well the system can do what it can do.** Help the user understand how often the AI system may make mistakes.	[Music Recommenders, Product #1] "A little bit of hedging language: 'we think you'll like.'"
During interaction	G3	**Time services based on context.** Time when to act or interrupt based on the user's current task and environment.	[Navigation, Product #1] "In my experience using the app, it seems to provide timely route guidance. Because the map updates regularly with your actual location, the guidance is timely."
	G4	**Show contextually relevant information.** Display information relevant to the user's current task and environment.	[Web Search, Product #2] "Searching a movie title returns show times near my location for today's date."
	G5	**Match relevant social norms.** Ensure the experience is delivered in a way that users would expect, given their social and cultural context.	[Voice Assistants, Product #1] "[The assistant] uses a semi-formal voice to talk to you—spells out 'okay' and asks further questions."
	G6	**Mitigate social biases.** Ensure the AI system's language and behaviors do not reinforce undesirable and unfair stereotypes and biases.	[Autocomplete, Product #2] "The autocomplete feature clearly suggests both genders [him, her] without any bias while suggesting the text to complete."
When wrong	G7	**Support efficient invocation.** Make it easy to invoke or request the AI system's services when needed.	[Voice Assistants, Product #1] "I can say [wake command] to initiate."
	G8	**Support efficient dismissal.** Make it easy to dismiss or ignore undesired AI system services.	[E-commerce, Product #2] "Feature is unobtrusive, below the fold, and easy to scroll past…Easy to ignore."
	G9	**Support efficient correction.** Make it easy to edit, refine, or recover when the AI system is wrong.	[Voice Assistants, Product #2] "Once my request for a reminder was processed I saw the ability to edit my reminder in the UI that was displayed. Small text underneath stated 'Tap to Edit' with a chevron indicating something would happen if I selected this text."
	G10	**Scope services when in doubt.** Engage in disambiguation or gracefully degrade the AI system's services when uncertain about a user's goals.	[Autocomplete, Product #1] "It usually provided 3–4 suggestions instead of directly auto completing it for you."
	G11	**Make clear why the system did what it did.** Enable the user to access an explanation of why the AI system behaved as it did.	[Navigation, Product #2] "The route chosen by the app was made based on the Fastest Route, which is shown in the subtext."
Over time	G12	**Remember recent interactions.** Maintain short term memory and allow the user to make efficient references to that memory.	[Web Search, Product #1] "[The search engine] remembers the context queries, with certain phrasing, so that it can continue the thread of the search (e.g., 'who is he married to' after a search that surfaces Benjamin Bratt)."
	G13	**Learn from user behavior.** Personalize the user's experience by learning from their actions over time.	[Music Recommenders, Product #2] "I think this is applied because every action to add a song to the list triggers new recommendations."
	G14	**Update and adapt cautiously.** Limit disruptive changes when updating and adapting the AI system's behaviors.	[Music Recommenders, Product #2] "Once we select a song they update the immediate song list below but keep the above one constant."
	G15	**Encourage granular feedback.** Enable the user to provide feedback indicating their preferences during regular interaction with the AI system.	[Email, Product #1] "The user can directly mark something as important, when the AI hadn't marked it as that previously."
	G16	**Convey the consequences of user actions.** Immediately update or convey how user actions will impact future behaviors of the AI system.	[Social Networks, Product #2] "[The product] communicates that hiding an Ad will adjust the relevance of future ads."
	G17	**Provide global controls.** Allow the user to globally customize what the AI system monitors and how it behaves.	[Photo Organizers, Product #1] "[The product] allows users to turn on their location history so the AI can group photos by where you have been."
	G18	**Notify users about changes.** Inform the user when the AI system adds or updates its capabilities.	[Navigation, Product #2] "[The product] does provide small in-app teaching callouts for important new features. New features that require my explicit attention are pop-ups."

SOURCE: Amershi et al., 2019. Used with permission.

more ethically than humans on the battlefield.[54] AI-enabled systems can detect patterns that humans miss and remove biases that result in flawed sampling and selection. AI systems can handle many more parameters and operate on faster time scales than humans can.[55] Alternatively, when circumstances or requirements of a human-AI team are not considered, inefficiencies can result. The Navy's littoral combat ship was designed to be highly automated and require a smaller crew. However, that smaller crew requires three times the training and significantly more-senior personnel. The ship requires only a third of the crew of a frigate and can only perform about a third of the missions that a frigate can achieve.[56] In this case, the human-AI teaming required significant additional training for the humans and limited the mission space in which the team could participate.

Other investments have emphasized human factors in developing AI systems in an effort to realize more of the potential of human-AI teams. Human-AI teaming for military purposes has become a significant investment in the United States and in China. DARPA is investing in a program called Artificial Social Intelligence for Successful Teams,[57] which is an effort to use observations of human partners and contextual clues as inputs to a machine-learning process that would classify human mental states. The AI partner of a team could then use that inferred state to offer context-specific information. The U.S. Army Research Laboratory is considering "robust but intelligent teaming," meaning that the contribution of the human teammate is varied, depending on physiological status and emotional state.[58] The Chi-

---

[54] Ronald C. Arkin, "Ethical Robots in War," *IEEE Technology and Society Magazine*, Vol. 28, No. 1, Spring 2009.

[55] Amandeep Singh Gill, "Artificial Intelligence and International Security: The Long View," *Ethics & International Affairs*, Vol. 33, No. 2, June 2019.

[56] Matthew Johnson and Alonso H. Vera, "No AI Is an Island: The Case for Teaming Intelligence," *AI Magazine*, Vol. 40, No. 1, 2019.

[57] Joshua Elliott, "Artificial Social Intelligence for Successful Teams (ASIST)," DARPA program information, undated.

[58] Robert K. Ackerman, "The Army's Newest Technology Is the Human Brain," *Signal*, August 1 2019.

nese New Generation Artificial Intelligence Development Plan,[59] announced in 2017, accounts for billions of dollars in research spending—notably, some human-AI teaming in military contexts. Speed and accuracy contribute to human acceptance of AI decisionmaking. China has the strategic advantage of a large, surveilled population as a source of training data and is rapidly following the United States in the development of AI targeting systems.[60]

Conclusions: Insights Across Applications

Use of AI associated with HPE requires customization and training to get a reliable or useful pairing. A system that can collect and analyze relevant data, developed for reliable and practical use, is an objective that will continue to be more aspiration than actuality for the foreseeable future.[61] Furthermore, as with all personalized products, the cost of fitting the system to the user must be considered as a part of the value of any application. Most likely, any mass-produced product will only work for a portion of potential users; the training and selection of users might be time-consuming and resource-intensive. The process of training and validating BCI must be optimized for practical application of technology for performance enhancement. As more (and more-personal) data are available, the opportunity for precision or individualized fitting is expected to increase the efficiency and insight provided by AI. And if fitting involves implantation, minimizing and managing associated risk will also be essential.

In developing human-AI teams, including IDSS, careful consideration must be given to human factors. Human understanding of and trust in the performance of IDSS limit the utility of AI support as much as actual biases or failed outcomes. For AI teaming to improve military efficiency, the systems have to be designed in a way that strategic decisionmaking is supported by the

[59] O'Meara, Sarah, "Will China Lead the World in AI by 2030?" *Nature*, Vol. 572, No. 7770, August 2018.

[60] Barnes and Chin, 2018.

[61] Anna Wexler, "Separating Neuroethics from Neurohype," *Nature: Biotechnology*, Vol. 16, No. 9, September 2018.

AI but not dependent on it alone.[62] When the AI fails, the human member of the team must be able to continue the task; alternatively, the AI should identify when human decisionmaking might be compromised. Studies from multiple fields and disciplines have articulated the superiority of AI when teamed with human operators, leveraging large amounts of data and the rapid computing ability of the machine together with the cognition and contextual understanding of the human mind. The United States, China, and Russia are the nations with the most work already done and greatest future potential; but Canada, France, Germany, India, Israel, Japan, South Korea, and the United Kingdom are reported to have capabilities in this area.[63]

---

[62] Colin M. Sattler, *Aviation Artificial Intelligence: How Will It Fare in the Multi-Domain Environment?* Fort Leavenworth, Kan.: School of Advanced Military Studies, U.S. Army Command and General Staff College, 2018.

[63] Gill, 2019.

CHAPTER FOUR

The Internet of Bodies

What Is the Internet of Bodies?

Increasingly, the Internet of Things (IoT)—the network of smart devices that
have become ubiquitous in everyday life—is becoming more intimate with
the human body and creating an unprecedented bodily reliance on Internet-
enabled devices. Some have referred to these technologies, and the data they
collect, as the Internet of Bodies (IoB).[1] These devices are used for a variety
of purposes, such as health care, improved efficiency, and enhanced human
performance.[2] Although there is no universal definition of this set of emerging technologies, for the purposes of this report, an *IoB device* is defined as

- a body-connected or medical device that
  - contains software or computing capabilities
  - can communicate with an Internet-connected device or network
  - satisfies one or more of the following:
    - measures or collects information about an individual's body
    - can alter the external environment or the human body's function-
      ality.[3]

---

[1] Meghan Neal, "The Internet of Bodies Is Coming, and You Could Get Hacked,"
Motherboard, March 13, 2014; Mary Lee, "The 'Internet of Bodies' Is Setting Dangerous
Precedents," *Washington Post*, October 15, 2018.

[2] Mary Lee, Benjamin Boudreaux, Ritika Chaturvedi, Sasha Romanosky, and Bryce
Downing, *The Internet of Bodies: Opportunities, Risks, and Governance*, Santa Monica,
Calif.: RAND Corporation, RR-3226-RC, 2020.

[3] IoB devices are restricted to technologies that can be linked to an individual person
rather than to traits that are more or less universal to all humans or to a particular dis-

A *body-connected device* is one whose usage and function requires either biofluids or physical contact with the body (e.g., is worn, ingested, implanted, or otherwise attached to or embedded in the body, temporarily or permanently). The definition for *medical device* is the same as that of the FDA.[4] Based on this definition, every IoB device is an IoT device.

The software or computing capabilities in an IoB device could be as simple as a few lines of code used to configure a microchip implant, or they could require a full computer that can run complex AI and machine-learning algorithms. The device need not directly connect to the Internet but could, for example, connect via Bluetooth to one's smartphone, which communicates (e.g., through a cellular or WiFi network) with a cloud server to send and receive relevant information to the user or external parties. Information about the human body that is collected by an IoB device might be health information (e.g., steps or heart rate). An alteration in the external environment refers to a physical change outside the human body, such as a smart door that unlocks by implanted microchip. An alteration to the body's functionality might be increased cognition promised by a BCI or the ability to record whatever the user sees through an intraocular lens equipped with a camera.

IoB devices can be consumer devices (i.e., devices meant for elective use) or medical devices (those prescribed by a health care provider or used in a health care system). They can generally be classified in the following ways:

---

ease. Therefore, large genetic sequence databases, such as GenBank, are not considered IoB data, and techniques that are enabled by such databases, such as CRISPR, are not considered IoB technologies.

[4] FDA, "How to Determine If Your Product Is a Medical Device," webpage, December 16, 2019. The FDA defines a *medical device* as:

- an instrument, apparatus, implement, machine, contrivance, implant, in vitro reagent, or other similar or related article, including a component part or accessory which is: recognized in the official National Formulary, or the United States Pharmacopoeia, or any supplement to them,
- intended for use in the diagnosis of disease or other conditions, or in the cure, mitigation, treatment, or prevention of disease, in man or other animals, or
- intended to affect the structure or any function of the body of man or other animals, and which does not achieve its primary intended purposes through chemical action within or on the body of man or other animals and which is not dependent upon being metabolized for the achievement of any of its primary intended purposes.

- wearables (e.g., fitness trackers, smart clothing, Bluetooth-enabled prosthetics, smart eyewear with augmented reality or virtual reality capabilities)
- ingestibles (smart or electronic pills)
- implantables (pacemakers, electronic tattoos, insulin pumps, radio frequency identification [RFID] implantable microchips, BCIs)
- freestanding devices (smart scales, smart beds, infusion pumps)
- databases (electronic health records, genetic testing kit databases)

Devices that do not fall within the IoB definition include ordinary heart rate monitors or medical ID bracelets that are not Internet-connected. Devices can fall in and out of the definition depending on their application. For example, an Internet-connected smartphone would not on its own be part of the IoB, but it would become part of the IoB if a health app were installed that required physical contact with the body to track steps, heart rate, or other data.

The IoB is a relatively new phenomenon and was predicted by the Institute of Electrical and Electronics Engineers as a top tech trend for 2019.[5] Because the IoT is already fraught with risks—cybersecurity, privacy, and ethical—the consensus is that the IoB will only exacerbate those problems, in addition to bringing about its own specific issues (which are further discussed in the section on IoB regulations and gaps). Experts agree that the IoB is already here and that adoption will continue to grow, either because of market demand or because it is becoming increasingly difficult to find devices that are not Internet-connected. Like other emerging technologies, the IoB is outpacing the rate at which policymakers can implement regulations that protect consumers and other stakeholders while encouraging innovation.

---

[5] Roberto Saracco, "2019 Tech Trends," *IEEE Future Directions Tech Blog*, December 22, 2018.

Evolution of the Internet of Bodies

Internet of Bodies Trends

The IoB market is expected to continue to grow, with technologies becoming more connected and increasingly intimate with the human body. Figure 4.1 illustrates some examples of the variety of IoB devices on the market or under development. Vast amounts of data about the individual will continue to be collected, which could enable improved health care and human performance.

Wearables are becoming increasingly advanced. Smart clothing, wearable video cameras, headsets meant for meditation or combating depression, and smart glasses are already on the market. Wearables are being

FIGURE 4.1
Internet of Bodies Examples

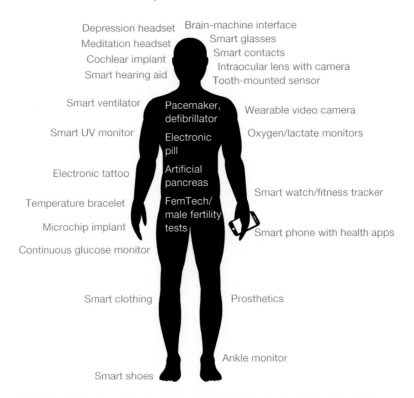

Depression headset
Meditation headset
Cochlear implant
Smart hearing aid
Smart ventilator
Smart UV monitor
Electronic tattoo
Temperature bracelet
Microchip implant
Continuous glucose monitor
Smart clothing
Smart shoes

Brain-machine interface
Smart glasses
Smart contacts
Intraocular lens with camera
Tooth-mounted sensor

Pacemaker, defibrillator
Electronic pill
Artificial pancreas
FemTech/ male fertility tests

Wearable video camera
Oxygen/lactate monitors
Smart watch/fitness tracker
Smart phone with health apps
Prosthetics
Ankle monitor

NOTE: FemTech = female technology (women's health technology); UV = ultraviolet.

developed to interact with the rest of the smart ecosystem; for example, smart clothing that measures the user's body temperature can automatically regulate a smart home thermostat. There are also wearable technologies that are meant for a certain sector of the market, such as FemTech devices that promise to improve women's health.[6]

The IoB will continue moving into the body, in the form of microchip implants,[7] electronic tattoos that can measure body temperature and connect to a smartphone by Bluetooth,[8] and tooth-mounted sensors that can keep track of food consumption.[9] Some people have implanted a combined router and hard drive device that can act as a node in a wireless mesh network.[10] Arguably the most advanced and invasive IoB technology under development is the BCI, which can both read and write to the brain. DARPA and commercial technology developers (such as Neuralink and Facebook) are working in this field.[11] Neuralink scientists have developed a "sewing machine" system for implanting electrodes in the brain, which will need clinical testing.[12] One goal of this technology is to restore lost physical function (e.g., to those who have lost a limb and can use a BCI to control a prosthetic). Other objectives are to improve human performance by making

---

[6] Gene Marks, "'Femtech' Startups on the Rise as Investors Scent Profits in Women's Health," *The Guardian*, June 6, 2019.

[7] Haley Weiss, "Why You're Probably Getting a Microchip Implant Someday," *The Atlantic*, September 21, 2018.

[8] Sieeka Khan, "Electronic Tattoos Can Be Made Through Graphene and Silk," *Science Times*, March 13, 2019.

[9] Mike Silver, "Scientists Develop Tiny Tooth-Mounted Sensors That Can Track What You Eat," *Tufts Now*, March 22, 2018.

[10] Daniel Oberhaus, "This DIY Implant Lets You Stream Movies from Inside Your Leg," *Wired*, August 30, 2019.

[11] Al Emondi, "Next-Generation Nonsurgical Neurotechnology," DARPA program information, undated; Neuralink, homepage, undated; Noam Cohen, "Zuckerberg Wants Facebook to Build a Mind-Reading Machine," *Wired*, March 7, 2019.

[12] Timothy L. Hanson, Camilo L. Diaz-Botia, Viktor Kharazia, Michael M. Maharbiz, and Philip N. Sabes, "The 'Sewing Machine' for Minimally Invasive Neural Recording," BioRxiv, March 14, 2019; Stephen Shankland, "Elon Musk Says Neuralink Plans 2020 Human Test of Brain-Computer Interface," CNET, July 17, 2019.

multitasking easier, enhancing cognition, or improving memory. More details about BCI are provided in Chapter Three.

Developments Related to the Internet of Bodies

Predictions have been made that advances in networking and connectivity such as 5G, 6G, and WiFi 6 (the fifth and sixth generations of telecommunications technology and the next generation of WiFi technology) will radically increase the ability of IoB and IoT devices to connect to each other and to the Internet. This is expected to result in more-personalized experiences and improved efficiencies. However, 5G, 6G, and WiFi 6 protocols all have verified security flaws.[13] (Note that 6G and WiFi 6 official standards are still under development.) The increased connectivity of IoT and IoB devices might also mean an increased attack surface, which will result in more vulnerabilities enabled by multiple new access points.

Related technologies might be enabled by the IoB in unexpected ways. An infrared laser has been developed that can detect a person's cardiac signature with over 95 percent accuracy from a distance of 200 meters, even through certain clothing.[14] Cardiac signatures are unique to each person, remain constant, and cannot be camouflaged or changed. This laser could, for example, be used for identification purposes if a database of cardiac signatures were available. It could also be used as a contactless way for doctors to scan for cardiac events or for hospitals to monitor patients. Thus, the application of the laser could be enabled by IoB data, even if the laser itself might not meet the IoB definition.

---

[13] Alfred Ng, "Security Flaw Allows for Spying over 5G, Researchers Warn," CNET, February 1, 2019; Patrick Nelson, "5G and 6G Wireless Technologies Have Security Issues," *Network World*, October 25, 2018; Dan Goodin, "Serious Flaws Leave WPA3 Vulnerable to Hacks That Steal Wi-Fi Passwords," *Ars Technica*, April 11, 2019.

[14] David Hambling, "The Pentagon Has a Laser That Can Identify People from a Distance—by Their Heartbeat," *MIT Technology Review*, June 27, 2019.

Major Players in Internet of Bodies Developments

The United States remains a leader in biotechnology.[15] China has long-term plans to become a major player in biotech, according to a 2019 report prepared for the U.S.-China Economic and Security Review Commission.[16] The United States overall is leading in spending for biotechnology research and development (R&D)—$30 billion in annual spending compared with China's $600 million[17]—but U.S. funding of life sciences R&D has been shrinking since 2010.[18] The Chinese government has created several initiatives—such as Made in China 2025, the Strategic Emerging Industries Initiative, and China's Five-Year Plans—to bring the country to the forefront in various technological advancements.[19] China's goal for the five-year period ending in 2020 was for its biotechnology sector to exceed 4 percent of its gross domestic product.[20] Foreign capital, in the forms of foreign direct investment (including greenfield FDI[21]) and venture capitalism, is a major mechanism that China aims to use to garner investments in Chinese tech

---

[15] Note that biotechnology has some overlap with the IoB, but there are technologies that fall into one category but not the other (e.g., environmental biotechnology is not IoB, and implanted microchips would likely not be considered biotechnology).

[16] Gryphon Scientific and Rhodium Group, *China's Biotechnology Development: The Role of US and Other Foreign Engagement*, U.S.-China Economic and Security Review Commission, February 14, 2019.

[17] Unless otherwise specified, all monetary amounts are in U.S. dollars.

[18] Gryphon Scientific and Rhodium Group, 2019. China's R&D spending is increasing, and since the biotech industry is growing, U.S. researchers might look to China for funding opportunities.

[19] Gryphon Scientific and Rhodium Group, 2019. China is on par with or becoming more competitive with the United States in some biotechnology subsectors (such as genomics, biologics, and CAR-T [chimeric antigen receptor T] cell therapy). Although these are not necessarily IoB technologies, these advancements demonstrate China's aptitude and ability to keep pace with research in the latest in technology development.

[20] Shannon Ellis, "Biotech Booms in China," *Nature*, Vol. 553, No. 7688, January 17, 2018b. For more aspirations for the current five-year plan, see Rolf D. Schmid and Xin Xiong, "Biotech in China 2021, at the Beginning of the 14th Five-Year Period ("145")," *Applied Microbiology and Biotechnology*, May 3, 2021.

[21] *Greenfield FDI* is a type of foreign direct investment in which a parent company creates a subsidiary company in another country.

companies, both to increase domestic competencies and to gain a foothold in new international markets.[22]

China also seeks to incentivize the return of Chinese researchers who go abroad for training. Various talent programs offer financial benefits, such as salaries, relocation costs, and startup funding. Many of these programs also recruit researchers from the United States and other countries to come to China.[23] A former secretary-general of one of these programs states that the life sciences committee for biotech is one of the largest groups in the program, and that more than 1,400 people from science and industry have been recruited.[24]

China is making an effort to build *biotechnology parks*—large campuses designed for colocation of tech companies around certain themes, such as nanotechnology.[25] The plan is to establish ten to 20 life-science parks for medicine with a combined output of at least $1.5 billion. It was reported in 2018 that more than 100 of these life-science parks can be found all over the country and that more than $100 billion has already been invested in the life-sciences sector by various state and local governments.[26] The purpose of these parks is to create innovation hubs and provide infrastructure, business support, and talent pools for multiple businesses in a given location.[27]

Chinese companies have been purchasing and licensing patents and technology from overseas in the past few years. This allows China the chance to catch up to advancements made in other countries; it also gives the licensee opportunities to exploit intellectual property. There have been several instances of intellectual property theft or data theft by China.[28]

China recognizes the importance of big-data applications to genomic and health care data, so it is investing in modernizing the collection and sharing of such data with goals of becoming a hub for data centers and cloud

---

[22] Gryphon Scientific and Rhodium Group, 2019.

[23] Gryphon Scientific and Rhodium Group, 2019.

[24] Ellis, 2018b.

[25] Gryphon Scientific and Rhodium Group, 2019.

[26] Ellis, 2018b.

[27] Gryphon Scientific and Rhodium Group, 2019.

[28] Gryphon Scientific and Rhodium Group, 2019.

computing. The Chinese government announced in fall 2018 that it planned to spend $15 billion over the next five years on these projects.[29] The United States is relatively less restrictive with the sharing of personal data domestically or internationally; in contrast, China has much stronger restrictions on sharing its citizens' data. The United States, therefore, might not have reciprocal access to health and genomic data.[30]

The U.S.-China Economic and Security Review Commission report also states that China's patents in biotechnology have increased rapidly between 1996 and 2016, and patents granted in China surpassed those granted in the United States in 2012 (see Figure 4.2). Biotechnology publications are growing at similar rates in the United States, China, and globally (Figure 4.3). China's progress is partly the result of research partnerships with U.S. institutions, bidirectional investments with the United States, and recruitment of foreign- and Chinese-born scientists who have been trained in the United States (see

FIGURE 4.2

Annual Biotechnology Patents Granted in the United States and China, 1996–2016

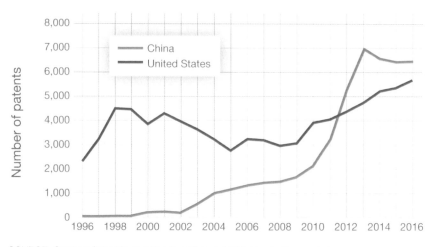

SOURCE: Gryphon Scientific and Rhodium Group, 2019. Used with permission.

---

[29] Yawen Chen, "China to Invest $15 Billion in Big Data, Cloud Computing over Next Five Years," Reuters, September 19, 2018.

[30] Gryphon Scientific and Rhodium Group, 2019.

FIGURE 4.3

Annual Biotechnology Publications by Country, 2000–2017

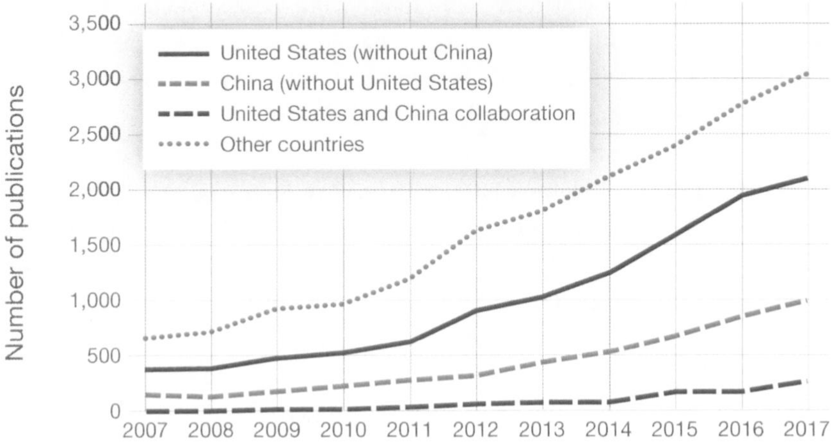

SOURCE: Gryphon Scientific and Rhodium Group, 2019. Used with permission.

Figure 4.4). Nonetheless, "continued investment by the US in its own biotechnology industry will ensure its dominance well into the future."[31]

Technology giants have entered IoB and related fields. Google is one partner in Project Baseline,[32] which conducts various longitudinal health studies and allows participants to test health wearables or other technologies. GV, the venture capital arm of Google's parent company Alphabet, is investing about one-third of its capital in health care and life science startups.[33] Apple is conducting a study with Stanford University on detecting heart rate irregularities with its Apple Watch.[34] One version of the watch also has FDA clearance to monitor signs of Parkinson's disease.[35] Amazon's cloud

---

[31] Gryphon Scientific and Rhodium Group, 2019.

[32] Project Baseline, homepage, undated.

[33] Sam Shead, "The VC Arm of Google's Parent Company Is Betting Its Billions on Life-Enhancing Healthcare Startups," *Business Insider,* September 10, 2017.

[34] "Apple Heart Study Demonstrates Ability of Wearable Technology to Detect Atrial Fibrillation," Stanford Medicine News Center, March 16, 2019.

[35] Daphne Allen, "Apple Gains FDA Clearance for ECG App for Apple Watch Series 4," Medical Device and Diagnostic Industry, September 13, 2018.

FIGURE 4.4
Annual Number of Overseas Chinese Students

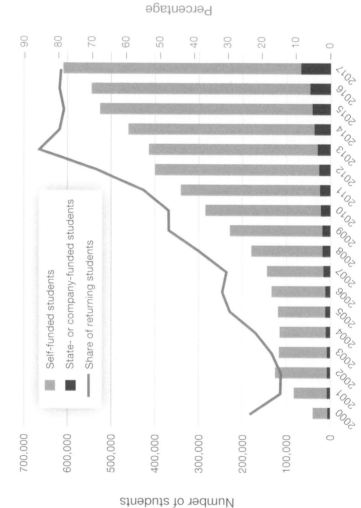

Percentage

Number of students

Self-funded students

State- or company-funded students

Share of returning students

SOURCE: Gryphon Scientific and Rhodium Group, 2019. Used with permission.

business, Amazon Web Services, has partnered with Merck and Accenture to develop a precision medicine platform.[36] Microsoft has partnered with 22 health companies to develop AI–machine-learning solutions.[37]

The United States still leads in medical device development.[38] Major players are the U.S. companies Medtronic (which moved its headquarters to Ireland for tax purposes[39]), DePuy Synthes, Thermo Fisher, and GE Health-care. The Dutch company Philips is another medical device leader.

In the wearable space, smartwatches, fitness trackers, and ear-worn devices have had the biggest market share.[40] U.S. technology developers, such as Fitbit and Apple, are leaders in the space, as are South Korean company Samsung and Chinese companies Xiaomi and Huawei. China gener-ated the most revenue in wearables in 2019, followed by the United States, India, the United Kingdom, and Germany.[41] In contrast, user penetration in 2019 is led by the United States, followed by Hong Kong, the United King-dom, China, and Australia.[42]

In addition to the United States and China, a few other countries have a foothold in IoB-related technologies. In a 2018 report by the Organisa-tion for Economic Cooperation and Development,[43] the United States was ranked as the top country in biotechnology patent applications, followed by the European Union, Japan, and Germany. That report also offers analysis

---

[36] Mike Miliard, "Accenture, Merck Partner with Amazon Web Services for New Cloud Precision Medicine Platform," *Healthcare IT News*, September 17, 2018.

[37] Microsoft, "Democratizing AI in Health," webpage, undated.

[38] Monique Ellis, "The Top 10 Medical Device Companies (2019)," *ProClinical*, blog post, May 29, 2019.

[39] Jeanne Whalen, "Medtronic's Ireland Move Results in Lower Taxes," *Wall Street Journal*, September 28, 2015.

[40] IDC, "IDC Reports Strong Growth in the Worldwide Wearables Market, Led by Hol-iday Shipments of Smartwatches, Wrist Bands, and Ear-Worn Devices," press release, March 5, 2019.

[41] Statista, "Wearables," webpage, undated.

[42] Statista, undated.

[43] Steffi Friedrichs, *Report on Statistics and Indicators of Biotechnology and Nanotech-nology*, Paris: OECD Science, Technology and Industry Working Papers, No. 2018/06, 2018.

of the top countries' Revealed Technology Advantage (RTA) index values. RTA is a measure of specialization in a field and is defined as the share of a country's granted patents in a field relative to the total share of that country's patents. Denmark, Australia, and New Zealand have had consistently high RTAs in biotechnology since 1990.

There are a few implantable microchip companies in the United States, notably Three Square Market and VivoKey.[44] Radio frequency identification (RFID) microchips are also popular in Sweden, where thousands of Swedes have had microchips (developed by Biohax International[45]) implanted for use with their metro system or to store emergency contact details.[46]

Outlook and Implications for the Internet of Bodies

Implications for Military and Intelligence Operations

The IoB could bring about national security consequences that neither users nor device makers intend. In early 2018, the fitness company Strava released a heat map of its users' running and exercise routes, which featured detailed geolocation data. The heat maps contained so much detail that they potentially exposed the locations and layouts of sensitive U.S. military bases.[47] The use of health tracking devices had been encouraged by the U.S. Department of Defense to help combat the obesity epidemic, and a pilot program had been conducted that gave out Fitbits to more than 2,000 soldiers in 2013 and 20,000 soldiers in 2015.[48] As a result of the Strava incident, the U.S.

---

[44] Three Square Market, homepage, undated; VivoKey, homepage, undated.

[45] F6S, "Biohax International: Turning the Internet of Things into the Internet of Us," webpage, undated.

[46] Maddy Savage, "Thousands of Swedes Are Inserting Microchips Under Their Skin," NPR, October 22, 2018.

[47] Alex Hern, "Fitness Tracking App Strava Gives Away Location of Secret US Army Bases," *The Guardian*, January 28, 2018.

[48] Amy Bushtaz, "Army Issues FitBit Bands in Test Fitness Program,"Military.com, October 22, 2013; Kevin Lilley, "20,000 Soldiers Tapped for Army Fitness Program's 2nd Trial," *Army Times*, July 27, 2015.

Department of Defense revised its policy and no longer permits deployed service members to use such apps or devices.[49]

There is also the national security risk of possible remote assassination of high-ranking U.S. officials through IoB devices. In 2007, Vice President Dick Cheney's doctors replaced his implanted heart defibrillator with one that was modified so that WiFi capabilities were disabled to prevent such an attack.[50] As IoB devices rise in popularity, an attack could be conceivable in a larger population and create a risk that affects not just U.S. government officials but a wide swath of ordinary civilians.

The Committee on Foreign Investment in the United States (CFIUS) examines foreign investment in and acquisition of U.S. companies for potential national security risks.[51] Kunlun, a Chinese gaming company, took control of the popular gay dating app Grindr in 2016. In May 2019, Kunlun agreed to sell Grindr following a CFIUS investigation.[52] Specific reasons for this change in ownership have not been disclosed by CFIUS, but it has been reported that blackmail of U.S. officials or government contractors is a national security concern because Grindr's databases contain sensitive information, such as its users' location data, messages, and HIV statuses.[53]

Furthermore, through investments and partnerships in U.S. health care companies, China has direct access to vast amounts of genetic and clinical data on U.S. residents. There are no reports or evidence that the Chinese government intends to use these data against Americans, but access to this information is considered a potential national security risk because of the possibility for blackmail.[54]

---

[49] Tara Copp, "Fitbits and Fitness-Tracking Devices Banned for Deployed Troops," *Military Times*, August 6, 2018.

[50] Dana Ford, "Cheney's Defibrillator Was Modified to Prevent Hacking," CNN, October 24, 2013.

[51] Echo Wang, "China's Kunlun Tech Agrees to U.S. Demand to Sell Grindr Gay Dating App," Reuters, May 13, 2019.

[52] Wang, 2019.

[53] Sarah Bauerle Danzman and Geoffrey Gertz, "Why Is the U.S. Forcing a Chinese Company to Sell the Gay Dating App Grindr?" *Washington Post*, April 3, 2019.

[54] Gryphon Scientific and Rhodium Group, 2019.

Regulations and Gaps

Although experts are trying to bring the IoB to the attention of lawmakers, various components of the IoB ecosystem are currently regulated by a patchwork of policies. Two U.S. federal laws apply to the protection of IoB data: the Health Insurance Portability and Accountability Act,[55] which applies only to covered entities (health providers, health plans, and health care clearinghouses) that work with medical data; and the Genetic Information Nondiscrimination Act,[56] which applies only to genetic information for health insurance or employment decisions. The European Union's General Data Protection Regulation (GDPR) is the main mechanism protecting citizens of the European Union in the event of IoB or other data breaches. GDPR outlines such regulations as a right to access data and a right to be forgotten. Because there is no comparable U.S. federal data privacy law, individual states have taken the lead in regulating personal information. For example, every state has a data breach notification law, but only some of them contain biometric data in their regulations, and biometrics are defined differently in each state. Illinois and Texas have implemented laws that require consent before biometric information is collected.[57] The California Consumer Privacy Act, effective January 2020, gives California residents more visibility regarding information that businesses are collecting about them, provides individuals with the right to opt out of allowing businesses to sell their data, and forbids discrimination or denial of access to service for those who do not provide their data. More states are starting to consider similar legislation.

The FDA regulates medical device safety in the United States and has been forward-looking regarding cybersecurity. For example, the agency has published pre- and postmarket cyber guidance documents for medical device manufacturers and has a *Digital Health Innovation Action Plan*.[58]

---

[55] U.S. Department of Health and Human Services, "Health Information Privacy," webpage, undated.

[56] Public Law 110-233, Genetic Information Nondiscrimination Act, 2008.

[57] Illinois Biometric Information Privacy Act, Public Law No. 740 ILCS 14/; Texas Business and Commerce Code, Title 11, Subtitle A, Chapter 503, 2009.

[58] FDA, *Digital Health Innovation Action Plan*, undated.

The FDA also partners with hackers and device makers to find and disclose vulnerabilities, hosts public workshops to solicit and revise guidance and best practices, and is working with the MITRE Corporation to develop a scoring system to rank cyber vulnerabilities in medical devices.[59]

However, the IoB presents several unanswered legal, policy, and security questions and potentially enables regulatory gaps and enforcement challenges. Law professor and cybersecurity expert Andrea Matwyshyn was the first scholar to apply the term "IoB" to law and policy discussions.[60] Matwyshyn has done the most work on the topic and has raised many unanswered legal questions relating to IoB.[61] One such question is regulatory purview—that is, whether certain IoB devices fall under FDA jurisdiction. Wrist-worn fitness trackers, such as Fitbits, are not considered medical devices and thus are not FDA-regulated; vitamin supplements are regulated, but the agency generally depends on manufacturers to verify their own products' safety. It is not obvious whether such a device as a hypothetical electronic pill that transmits data about gut health would meet the FDA's medical device definition; if it did not, the manufacturer would not be required to meet the agency's patient safety guidelines. Regulation would fall instead to the Federal Trade Commission, which is a much smaller agency with fewer resources for enforcement. Another legal issue that Matwyshyn brings up is contract law. End-user license agreements allow device makers to retain software rights and monitor and share users' data. Some of these agreements can even allow companies to "brick" (deactivate) a device unless a user has agreed to a new privacy or information-sharing policy, which could have major bodily implications for a device implanted in the user. There is also the issue of patent law, which might result in an end to manufacturer support (e.g., patches of vulnerabilities) of the device if a patent infringement has been discovered. Matwyshyn also brings up the issue of bankruptcy: If an IoB device maker goes bankrupt,

---

[59] Penny Chase and Steve Christey Coley, *Rubric for Applying CVSS to Medical Devices*, Washington, D.C.: MITRE, January 16, 2019.

[60] Andrea M. Matwyshyn, homepage, undated.

[61] See, for example, "Cyber Risk Wednesday: Internet of Bodies" webcast, Atlantic Council, September 21, 2017; and Andrea Matwyshyn, "The 'Internet of Bodies' Is Here. Are Courts and Regulators Ready?" *Wall Street Journal*, November 12, 2018.

other entities could purchase the company's data, including personally identifiable consumer information.

Other experts have brought up questions specifically related to networked medical and health care devices. A 2015 report by the Atlantic Council outlines benefits, such as improved health outcomes and quality of life;[62] risks, such as privacy issues and malfunction because of cyberattack; and recommendations for improved safety, such as building security into the devices from the beginning of development and improving public-private and private-private collaborations. Some experts have written commentaries warning about ethical, privacy, and cyber risks of related topics, such as "our cyborg future,"[63] smart pills,[64] or the combination of the IoB with AI.[65]

Another potential risk brought on by IoB is that if a device requires a critical software update, a user might face issues if he or she relies on the device for a vital health care need but is out of network and unable to install the patch. In addition, data policies for information that falls outside the purview of the Health Insurance Portability and Accountability Act are murky. Data storage and risk of breach are concerns, but strict storage specifications must be balanced with users' desires to have easy access to their own data. Another question is data ownership. It is not clear who owns IoB data—the user, the device maker, or the health care practitioner. Many device makers sell their customers' data to third-party data brokers, whose practices are not transparent and are largely unregulated. Although law-

---

[62] Jason Healey, Neal Pollard, and Beau Woods, *The Healthcare Internet of Things: Risks and Rewards*, Washington, D.C.: Atlantic Council, March 2015.

[63] Benjamin Wittes and Jane Chong, *Our Cyborg Future: Law and Policy Implications*, Washington, D.C.: Brookings Institution, September 2014.

[64] Robert Klitzman, "'Smart' Pills Are Here and We Need to Consider the Risks," CNN, March 17, 2019.

[65] Nicole Lindsey, "Internet of Bodies: The Privacy and Security Implications," *CPO Magazine*, December 14, 2018.

makers and experts have been calling for data broker laws,[66] Vermont was the only state as of 2020 to have any such regulations.[67]

Serious questions arise regarding who has access to personal data and how those data are used. For instance, health insurers might have already used information (such as hobbies, purchasing habits, and social media content) that reveals unhealthy lifestyles to raise insurance rates.[68] China has been using DNA data, obtained from U.S. researchers, to surveil Uighurs.[69] Some people have questioned whether access to certain IoB data is an infringement on Fourth and Fifth Amendment rights. For example, law enforcement officers have been able to arrest suspects using pacemaker data and genetic genealogy (genetics and ancestry) data.[70] Data endurance is another concern. For instance, if genetic tests show that a person is a carrier of a disease that might get passed down to her children, can those children one day be denied certain insurance (e.g., disability or life insurance) or other benefits?

These are all issues that will likely require regulation and enforcement by appropriate agencies with sufficient resources. In the United States, proposals have been introduced in Congress to regulate IoT devices and data brokers. The Internet of Things Cybersecurity Improvement Act became law in December 2020 and mandates cybersecurity standards (to be developed by the National Institute of Standards and Technology) for IoT devices used by the federal government. IoB development has already outpaced the rate at which decisionmakers can implement policies to protect relevant stakeholders. Policymakers will need to strike the right balance to encourage innovation while protecting all stakeholders involved.

---

[66] "State AGs Call for Data Regulations," document reprint, *Washington Post*, June 24, 2019.

[67] Vermont Public Law No. H.764, Act 171, An Act Relating to Data Brokers and Consumer Protection, 2019.

[68] Marshall Allen, "Health Insurers Are Vacuuming Up Details About You—And It Could Raise Your Rates," *ProPublica*, July 17, 2018.

[69] Sui-Lee Wee, "China Uses DNA to Track Its People, with the Help of American Expertise," *New York Times*, February 21, 2019.

[70] Cleve R. Wootson, "A Man Detailed His Escape from a Burning House. His Pacemaker Told Police a Different Story," *Washington Post*, February 8, 2017; Eric Levenson, "It Started as a Hobby. Now They're Using DNA to Help Cops Crack Cold Cases," CNN, March 27, 2019.

CHAPTER FIVE

Conclusion

The three technology categories considered in this report have the potential to be used to enhance human performance, but at different points in time. A much-hyped area is gene manipulation, which is the most nascent—research focuses on the potential to alleviate disease, and decades of research will be required before meaningful application to HPE should be expected. That said, some forms of genetic analysis can be helpful in understanding capabilities of people as they are today. AI also has long been associated with hype, and although related technologies are steadily improving, meaningful impacts on cognitive and other prostheses are expected to take five to ten years. As with IOB devices, these new prostheses will provide new channels of communication, along with associated vulnerabilities. Finally, the IoB is already having an effect today; it will become more complex and capable over time. It is becoming both a source of signals and a target for cyberattacks.

All of these technologies, even the most aspirational, have implications for military and intelligence activities for the United States, its allies, and its adversaries. Although meaningful impact on HPE might not be expected for several years, the consequences when it is realized will be many and varied. This is one reason that, within the U.S. Department of Defense, the Under Secretary for Research and Engineering added an Assistant Director for biotechnology in 2019.[1]

Government initiatives around the world have focused on biotechnology and the life sciences, thanks in part to the maturation of technologies for analyzing data. As discussed, China has made biotechnology a national

---

[1] John Cumbers, "'With Great Power Comes Great Responsibility'—Meet Alexander Titus, the Department of Defense's Head of Biotechnology," *Forbes*, September 24, 2019.

priority, but it is not alone. The military laboratories of the former Soviet Union slowly converted to nonmilitary research areas and provided a formidable biotechnology base. Many of the best scientists left the country or retired over the course of the 1990s. Russia's biotechnology sector stalled in the early 2000s, but Russian government bolstered the sector in 2012 with a national plan called Biotec2020.[2] Western sanctions following Russia's seizure of Crimea and eastern Ukraine motivated efforts aimed at advancing the independence of Russia's biotechnology sector and forming partnerships with other nations, importantly China. Russia's sovereign wealth fund has invested in partnerships with leading Chinese biotechnology and genomic companies.[3] The private sector plays a key role in what is now discussed as the bioeconomy.[4]

Understanding when and where breakthroughs or setbacks have occurred—and when and where novel and even surprising applications have emerged—will require monitoring. Relevant program launches or terminations, awards of government grants or contracts, joint ventures and acquisitions, conferences and trade shows, patents, and publications are among the indicators to watch. It will also be important to monitor the kinds of data that are intentionally and unintentionally collected, used, or emitted by HPE-related systems. The benefits of emerging forms of communications will be offset by new kinds of cyber vulnerabilities.

---

[2] The national initiative set agrobio and food priority as areas of focus.

[3] Alina Osmakova, Michael Kirpichnikov, and Vladimir Popov, "Recent Biotechnology Developments and Trends in the Russian Federation," *New Biotechnology*, Vol. 40, Part A, January 25, 2019; Stanislav Tkachenko, "Future Complicated for Russian Biotech," *Russia Beyond*, October 30, 2014; Sintia Radu, "Russia and China's High-Tech Bet," *U.S. News and World Report*, November 29, 2019; Samuel Bendett and Elsa Kania, "A New Sino-Russian High-Tech Partnership," Australian Strategic Policy Institute, October 29, 2019.

[4] The White House cast its own spotlight on the bioeconomy. See Office of Science and Technology Policy, *Summary of the 2019 White House Summit on America's Bioeconomy*, White House, October 2019.

Abbreviations

AI	artificial intelligence
BCI	brain-computer interface
CFTR	cystic fibrosis transmembrane conductance regulator
CRISPR	clustered regularly interspaced short palindromic repeat
DARPA	Defense Advanced Research Projects Agency
DNA	deoxyribonucleic acid
EEG	electroencephalography
FDA	U.S. Food and Drug Administration
FemTech	female technology
fMRI	functional magnetic resonance imaging
GWAS	genome-wide association studies
HPE	human-performance enhancement
IDSS	intelligent decision support systems
IoB	Internet of Bodies
IoT	Internet of Things
MEG	magnetoencephalography
PET	positron emission tomography
SNP	single nucleotide polymorphism
XAI	explainable artificial intelligence

Bibliography

Ackerman, Robert K., "The Army's Newest Technology Is the Human Brain," *Signal,* August 1, 2019.

Allen, Daphne, "Apple Gains FDA Clearance for ECG App for Apple Watch Series 4," Medical Device and Diagnostic Industry, September 13, 2018.

Allen, Marshall, "Health Insurers Are Vacuuming Up Details About You—And It Could Raise Your Rates," *ProPublica*, July 17, 2018.

Allenby, Braden, "Designer Warriors: Altering Conflict—and Humanity Itself?" *Bulletin of the Atomic Scientists*, Vol. 74, No. 6, 2018.

Alliance for Regenerative Medicine, *Quarterly Regenerative Global Data Report: Q1 2019*, Washington, D.C., 2019.

Alton, Eric W. F. W., David K. Armstrong, Deborah Ashby, Katie J. Bayfield, Diana Bilton, Emily V. Bloomfield, A. Christopher Boyd, June Brand, Ruaridh Buchan, Roberto Calcedo, et al., "Repeated Nebulisation of Non-Viral CFTR Gene Therapy in Patients with Cystic Fibrosis: A Randomised, Double-Blind, Placebo-Controlled, Phase 2b Trial," *Lancet Respiratory Medicine,* Vol. 3, No. 9, September 2015, pp. 684–691.

American National Standards Institute and Association for the Advancement of Medical Instrumentation, *Human Factors Engineering—Design of Medical Devices*, New York, preview edition, 2013.

Amershi, Saleema, Dan Weld, Mihaela Vorvoreanu, Mihaela Vorvoreanu, Adam Fourney, Besmira Nushi, Penny Collisson, Jina Suh, Shamsi T. Iqbal, Paul N. Bennett, Kori Marie Inkpen, et al., "Guidelines for Human-AI Interaction," *Proceedings of the 2019 CHI Conference on Human Factors in Computing Systems*, Glasgow, Scotland, 2019.

Anumanchipalli, Gopala Krishna, Josh Chartier, and Edward F. Chang, "Speech Synthesis from Neural Decoding of Spoken Sentences," *Nature,* Vol. 568, No. 7753, 2019, pp. 493–498.

"Apple Heart Study Demonstrates Ability of Wearable Technology to Detect Atrial Fibrillation," Stanford Medicine News Center, March 16, 2019.

Arkin, Ronald C., "Ethical Robots in War," *IEEE Technology and Society Magazine*, Vol. 28, No. 1, Spring 2009, pp. 30–33.

Army Science Board, *Army Efforts to Enhance Soldier and Team Performance*, Washington, D.C., 2017.

Atkinson, Robert D., and Caleb Foote, *Is China Catching Up to the United States in Innovation?* Washington, D.C.: Information Technology & Innovation Foundation, April 8, 2019.

Banerjee, Sunayana B., Filomene G. Morrison, and Kerry J. Ressler, "Genetic Approaches for the Study of PTSD: Advances and Challenges," *Neuroscience Letters*, Vol. 649, 2017, pp. 139–146.

Bansal, Gagan, Besmira Nushi, Ece Kamar, Daniel S. Weld, Walter S. Lasecki, and Eric Horvitz, "Updates in Human-AI Teams: Understanding and Addressing the Performance/Compatibility Tradeoff," *Proceedings of the AAAI Conference on Artificial Intelligence*, Vol. 33, No. 1, July 17, 2019.

Barnes, Julian E., and Josh Chin, "The New Arms Race in AI," *Wall Street Journal*, March 2, 2018.

Bauerle Danzman, Sarah, and Geoffrey Gertz, "Why Is the U.S. Forcing a Chinese Company to Sell the Gay Dating App Grindr?" *Washington Post*, April 3, 2019.

Bendett, Samuel, and Elsa Kania, "A New Sino-Russian High-Tech Partnership," Australian Strategic Policy Institute, October 29, 2019.

Bini, Stefano A., "Artificial Intelligence, Machine Learning, Deep Learning, and Cognitive Computing: What Do These Terms Mean and How Will They Impact Health Care?" *Journal of Arthroplasty*, Vol. 33, No. 8, August 2018, pp. 2358–2361.

Binnendijk, Anika, Tim Marler, and Elizabeth M. Bartels, *Brain-Computer Interfaces: U.S. Military Applications and Implications, An Initial Assessment*, Santa Monica, Calif.: RAND Corporation, RR-2996-RC, 2020. As of August 16, 2021:
https://www.rand.org/pubs/research_reports/RR2996.html

van den Bosch, Karel, and Adelbert Bronkhorst, "Human-AI Cooperation to Benefit Military Decision Making," *Proceedings of the NATO IST-160 Specialist Meeting on Big Data and Artificial Intelligence for Military Decision Making*, Bordeaux, France, 2018.

"Brain-Controlled System Neurochat Begins to Be Batch-Produced in Russia," TASS Russian News Agency, April 23, 2019.

Breznitz, Dan, and Michael Murphree, *The Rise of China in Technology Standards: New Norms in Old Institutions*, research report prepared on behalf of the U.S.-China Economic and Security Review Commission, January 16, 2013.

British Medical Association, *Biotechnology, Weapons and Humanity*, London, Harwood Academic Publishers, 1999.

Bush, William S., and Jason H. Moore, "Chapter 11: Genome-Wide Association Studies," *PLOS Computational Biology*, Vol. 8, No. 12, 2012.

Bushtaz, Amy, "Army Issues FitBit Bands in Test Fitness Program," Military. com, October 22, 2013. As of July 20, 2020: https://www.military.com/daily-news/2013/10/22/ army-issues-fitbit-bands-in-test-fitness-program.html

Carmody, Rachel N., and Aaron L. Baggish, "Working Out the Bugs: Microbial Modulation of Athletic Performance," *Nature Metabolism*, Vol. 1, 2019, pp. 658–659.

Case, Nicky, "How to Become a Centaur," *Journal of Design and Science*, January 8, 2018.

Centre for Bioelectric Interfaces, homepage, undated. As of July 27, 2020: https://bioelectric.hse.ru/en/

Charlet, Katherine, "The New Killer Pathogens: Countering the Coming Bioweapons Threat," *Foreign Affairs*, Vol. 97, No. 3, 2018, pp. 178–184.

Chase, Penny, and Steve Christey Coley, *Rubric for Applying CVSS to Medical Devices*, Washington, D.C.: MITRE, January 16, 2019.

Chaudhary, Ujwal, Niels Birbaumer, and Ander Ramos-Murguialday, "Brain– Computer Interfaces for Communication and Rehabilitation," *Nature Reviews Neurology*, Vol. 12, 2016.

Chen, Xiaogang, Bing Zhao, Yijun Wang, Shengpu Xu, and Xiaorong Gao, "Control of a 7-DOF Robotic Arm System with an SSVEP-Based BCI," *International Journal of Neural Systems*, Vol. 28, No. 8, 2018.

Chen, Yawen, "China to Invest $15 Billion in Big Data, Cloud Computing over Next Five Years," Reuters, September 19, 2018.

Cheng, Lefeng, and T. Yu, "A New Generation of AI: A Review and Perspective on Machine Learning Technologies Applied to Smart Energy and Electric Power Systems," *International Journal of Energy Research*, Vol. 43, No. 6, 2019, pp. 1928–1973.

"China Unveils Brain-Computer Interface Chip," Xinhua Net, May 18, 2019. As of July 27, 2020: http://www.xinhuanet.com/english/2019-05/18/c_138069590.htm

Chinese Clinical Trial Registry, database, undated. As of July 12, 2019: http://www.chictr.org.cn/enindex.aspx

Chui, Michael, James Manyika, Mehdi Miremadi, Nicolaus Henke, Rita Chung, Pieter Nel, and Sankalp Malhotra, *Notes from the AI Frontier: Insights from Hundreds of Use Cases*, McKinsey Global Institute, 2018.

Clarivate, Web of Science database, undated. As of June 10, 2019: http://apps.webofknowledge.com/

Clayton, Ellen Wright, "A Genetically Augmented Future," *Nature*, Vol. 564, No. 7735, December 13, 2018.

Cohen, Jon, "The Untold Story of the 'Circle of Trust' Behind the World's First Gene-Edited Babies," *Science*, August 1, 2019.

Cohen, Noam, "Zuckerberg Wants Facebook to Build a Mind-Reading Machine," *Wired*, March 7, 2019.

Colzato, Lorenza S., Michael A. Nitsche, and Armin Kibele, "Noninvasive Brain Stimulation and Neural Entrainment Enhance Athletic Performance—A Review," *Journal of Cognitive Enhancement*, Vol. 1, No. 1, 2017, pp. 73–79.

Convention on the Prohibition of the Development, Production and Stockpiling of Bacteriological (Biological) and Toxin Weapons and on Their Destruction, disarmament treaty, entered into force March 26, 1975.

Copp, Tara, "Fitbits and Fitness-Tracking Devices Banned for Deployed Troops," *Military Times*, August 6, 2018.

Cornelis, Marilyn C., Nicole R. Nugent, Ananda B. Amstadter, and Karestan C. Koenen, "Genetics of Post-Traumatic Stress Disorder: Review and Recommendations for Genome-Wide Association Studies," *Current Psychiatry Reports*, Vol. 12, No. 4, 2010, pp. 313–326.

Cumbers, John, "'With Great Power Comes Great Responsibility'—Meet Alexander Titus, the Department of Defense's Head of Biotechnology," *Forbes*, September 24, 2019.

"Cyber Risk Wednesday: Internet of Bodies" webcast, Atlantic Council, September 21, 2017.

Cyranoski, David, "CRISPR-Baby Scientist Fails to Satisfy Critics," *Nature*, Vol. 564, No. 7734, 2018, pp. 13–14.

Cyranoski, David, "What CRISPR-Baby Prison Sentences Mean for Research: Chinese Court Sends Strong Signal by Punishing He Jiankui and Two Colleagues," *Nature*, Vol. 577, 2020.

Daws, Ryan, "Putin Outlines Russia's National AI Strategy Priorities," *AI News*, May 31, 2019.

Douam, Florian, Jenna M. Gaska, Benjamin Y. Winer, Qiang Ding, Markus von Schaewen, and Alexander Ploss, "Genetic Dissection of the Host Tropism of Human-Tropic Pathogens," *Annual Review of Genetics*, Vol. 49, 2015, pp. 21–45.

Dunbar, Cynthia E., Katherine A. High, J. Keith Joung, Donald B. Kohn, Keiya Ozawa, and Michel Sadelain, "Gene Therapy Comes of Age," *Science*, Vol. 359, No. 6372, January 2018.

Elish, Madeline Claire, "Moral Crumple Zones: Cautionary Tales in Human-Robot Interaction," *Engaging Science Technology and Society*, Vol. 5, 2019, pp. 40–60.

Elliott, Joshua, "Artificial Social Intelligence for Successful Teams (ASIST)," DARPA program information, undated. As of July 20, 2020:
https://www.darpa.mil/program/
artificial-social-intelligence-for-successful-teams

Ellis, Monique, "The Top 10 Medical Device Companies (2019)," *ProClinical*, blog post, May 29, 2019. As of July 20, 2020:
https://www.proclinical.com/blogs/2019-5/
the-top-10-medical-device-companies-2019

Ellis, Shannon, "China's Fledgling Biotech Sector Fizzes into Life," *Nature Biotechnology*, Vol. 36, 2018a, pp. 8–9.

Ellis, Shannon, "Biotech Booms in China," *Nature*, Vol. 553, No. 7688, January 17, 2018b.

Emondi, Al, "Next-Generation Nonsurgical Neurotechnology," DARPA program information, undated. As of July 20, 2020:
https://www.darpa.mil/program/
next-generation-nonsurgical-neurotechnology

Eynon, Nir, Erik D. Hanson, Alejandro Lucia, Peter J. Houweling, Fleur C. Garton, Kathryn N. North, and David John Bishop, "Genes for Elite Power and Sprint Performance: ACTN3 Leads the Way," *Sports Medicine*, Vol. 43, No. 9, 2013, pp. 803–817.

F6S, "Biohax International: Turning the Internet of Things into the Internet of Us," webpage, undated. As of July 20, 2020:
https://www.f6s.com/biohaxinternational

FDA—*See* U.S. Food and Drug Administration.

Ford, Dana, "Cheney's Defibrillator Was Modified to Prevent Hacking," CNN, October 24, 2013.

Friedrichs, Steffi, *Report on Statistics and Indicators of Biotechnology and Nanotechnology*, Paris: OECD Science, Technology and Industry Working Papers, No. 2018/06, 2018.

"The Future Lies in Automation," *The Economist*, April 6, 2017.

Geist, Edward, and Andrew J. Lohn, *How Might Artificial Intelligence Affect the Risk of Nuclear War?* Santa Monica, Calif.: RAND Corporation, PE-296-RC, 2018. As of July 20, 2020:
https://www.rand.org/pubs/perspectives/PE296.html

Georgiades, Evelina, Vassilis Klissouras, Jamie Baulch, Guan Wang, and Yannis Pitsiladis, "Why Nature Prevails over Nurture in the Making of the Elite Athlete," *BMC Genomics*, Vol. 18, Suppl. 8, 2017.

Gerstein, Daniel, and James Giordano, "Rethinking the Biological and Toxin Weapons Convention?" *Health Security*, Vol. 15, No. 6, 2017.

Gill, Amandeep Singh, "Artificial Intelligence and International Security: The Long View," *Ethics & International Affairs*, Vol. 33, No. 2, June 2019, pp. 169–179.

Ginn, Samantha L., Anais K. Amaya, Ian E. Alexander, Michael Edelstein, and Mohammad R. Abedi, "Gene Therapy Clinical Trials Worldwide to 2017: An Update," *Journal of Gene Medicine*, Vol. 20, No. 5, 2018.

Goebel, Randy, Ajay Chander, Katharina Holzinger, Freddy Lecue, Zeynep Akata, Simone Stumpf, Peter Kieseberg, and Andreas Holzinger, "Explainable AI: The New 42?" in Andreas Holzinger, Peter Kieseberg, A Min Tjoa, and Edgar Weippl, eds., *Machine Learning and Knowledge Extraction, Cd-Make 2018*, Vol. 11015, August 2018, pp. 295–303.

Goh, R. Y., and Lai Soon Lee, "Credit Scoring: A Review on Support Vector Machines and Metaheuristic Approaches," *Advances in Operations Research*, Vol. 2019, No. 2, March 2019, pp. 1–30.

Gonfalonieri, Alexandre, "A Beginner's Guide to Brain-Computer Interface and Convolutional Neural Networks," *Towards Data Science*, November 25, 2018.

Goodin, Dan, "Serious Flaws Leave WPA3 Vulnerable to Hacks That Steal Wi-Fi Passwords," *Ars Technica*, April 11, 2019.

Goodwin, Kate, "China's CRISPR Babies Propel WHO to Issue Global Standards for Gene Editing," *BioSpace*, July 13, 2021.

Greene, Marsha, and Zubin Master, "Ethical Issues of Using CRISPR Technologies for Research on Military Enhancement," *Journal of Bioethical Inquiry*, Vol. 15, No. 3, 2018.

Gryphon Scientific and Rhodium Group, *China's Biotechnology Development: The Role of US and Other Foreign Engagement*, U.S.-China Economic and Security Review Commission, Feb 14, 2019. As of September 16, 2021: https://www.gryphonscientific.com/wp-content/uploads/2019/07/China%E2%80%99s-Biotechnology-Development-The-Role-of-U.S.-and-Other-Foreign-Engagement.pdf

Hambling, David, "The Pentagon Has a Laser That Can Identify People from a Distance—by Their Heartbeat," *MIT Technology Review*, June 27, 2019.

Hanson, Timothy L., Camilo L. Diaz-Botia, Viktor Kharazia, Michael M. Maharbiz, and Philip N. Sabes, "The 'Sewing Machine' for Minimally Invasive Neural Recording," *BioRxiv*, March 14, 2019.

Harvard Medical School Consortium for Space Genetics, "About Us," webpage, undated. As of June 16, 2019: https://spacegenetics.hms.harvard.edu/about-us

Headrick, Dan, and MaryAnne M. Gobble, "AI-Powered Fintech Turns Data into New Business," *Research-Technology Management*, Vol. 62, No. 1, January–February 2019, pp. 5–7.

Healey, Jason, Neal Pollard, and Beau Woods, *The Healthcare Internet of Things: Risks and Rewards*, Washington, D.C.: Atlantic Council, March 2015.

Hern, Alex, "Fitness Tracking App Strava Gives Away Location of Secret US Army Bases," *The Guardian*, January 28, 2018.

Ho, Calvin W. L., Derek Soon, Karel Caals, and Jeevesh Kapur, "Governance of Automated Image Analysis and Artificial Intelligence Analytics in Healthcare." *Clinical Radiology*, Vol. 74, No. 5, March 2019, pp. 329–337.

Hoffman, Robert R., Gary Klein, and Shane T. Mueller, "Explaining Explanation for 'Explainable AI,'" *Proceedings of the Human Factors and Ergonomics Society Annual Meeting*, September 25, 2018.

Holzinger, Andreas, Peter Kieseberg, Edgar Weippl, and A. Min Tjoa, "Current Advances, Trends and Challenges of Machine Learning and Knowledge Extraction: From Machine Learning to Explainable AI," paper presented at International Cross-Domain Conference, Hamburg, Germany, August 2018.

Horvath, Jared C., Olivia Carter, and Jason D. Forte, "Transcranial Direct Current Stimulation: Five Important Issues We Aren't Discussing (But Probably Should Be)," *Frontiers in Systems Neuroscience*, Vol. 8, No. 2, January 2014.

Hramov, Alexander E., Vladimir A. Maksimenko, and Marina Hramova, "Brain-Computer Interface for Alertness Estimation and Improving," *Dynamics and Fluctuations in Biomedical Photonics XV Proceedings*, Vol. 10493, May 21, 2018.

Hramov, Alexander E., Vladimir A. Maksimenko, Maxim D. Zhuravlev, and Alexander N. Pisarchik, "Immediate Effect of Neurofeedback in Passive BCI for Alertness Control," paper presented at the 7th International Winter Conference on Brain-Computer Interface (BCI), High 1 Resort, Korea, February 2019.

Huang, Futao, "Quality Deficit Belies the Hype: Few Chinese Researchers Are Regarded as Global Leaders, as the Pressure for Rapid Output Prevails," *Nature*, Vol. 564, No. 7735, December 2018, pp. S70–S71.

Huang, Qingyang, "Genetic Study of Complex Diseases in the Post-GWAS Era," *Journal of Genetics and Genomics*, Vol. 42, No. 3, 2015, pp. 87–98.

Ibrahim-Verbaas, Carla A., J. Bressler, Stéphanie Debette, Maaike Schuur, A. V. Smith, J. C. Bis, Gail Davies, Stella Trompet, J. A. Smith, C. Wolf, et al., "GWAS for Executive Function and Processing Speed Suggests Involvement of the CADM2 Gene," *Molecular Psychiatry*, Vol. 21, No. 2, 2016, pp. 189–197.

IDC, "IDC Reports Strong Growth in the Worldwide Wearables Market, Led by Holiday Shipments of Smartwatches, Wrist Bands, and Ear-Worn Devices," press release, March 5, 2019.

Ikegawa, Shiro, "A Short History of the Genome-Wide Association Study: Where We Were and Where We Are Going," *Genomics & Informatics*, Vol. 10, No. 4, 2012, pp. 220–225.

Ilardo, Melissa A., Ida Moltke, Thorfinn S. Korneliussen, Jade Cheng, Aaron J. Stern, Fernando Racimo, Peter de Barros Damgaard, Martin Sikora, Andaine Seguin-Orlando, Simon Rasmussen, et al., "Physiological and Genetic Adaptations to Diving in Sea Nomads," *Cell*, Vol. 173, No. 3, 2018, pp. 569–580.

Ilardo, Melissa, and Rasmus Nielsen, "Human Adaptation to Extreme Environmental Conditions," *Current Opinion in Genetics & Development*, Vol. 53, December 2018, pp. 77–82.

Illinois Biometric Information Privacy Act, Public Law No. 740 ILCS 14/.

Jackson, Maria, Leah Marks, Gerhard H. W. May, and Joanna B. Wilson, "The Genetic Basis of Disease," *Essays in Biochemistry*, Vol. 62, No. 5, 2018, pp. 643–723.

Jansen, Phillip R., Kyoko Watanabe, Sven Stringer, Nathan Skene, Julien Bryois, Anke R Hammerschlag, Christiaan A de Leeuw, Jeroen S Benjamins, Ana B Muñoz-Manchado, Mats Nagel, et al., "Genome-Wide Analysis of Insomnia in 1,331,010 Individuals Identifies New Risk Loci and Functional Pathways," *Nature Genetics*, Vol. 51, No. 3, 2019, pp. 394–403.

Johnson, Matthew, and Alonso H. Vera, "No AI Is an Island: The Case for Teaming Intelligence," *AI Magazine*, Vol. 40, No. 1, 2019, pp. 16–28.

Kania, Elsa B., "Minds at War: China's Pursuit of Military Advantage Through Cognitive Science and Biotechnology," *PRISM*, Vol. 3, No. 8, 2020.

Kashif, Malik, and Anterpreet Dua, "Biofeedback," StatPearls, December 20, 2019. As of May 6, 2020:
https://www.ncbi.nlm.nih.gov/books/NBK553075/

Khan, Sieeka, "Electronic Tattoos Can Be Made Through Graphene and Silk," *Science Times*, March 13, 2019.

Klitzman, Robert, "'Smart' Pills Are Here and We Need to Consider the Risks," CNN, March 17, 2019.

Knight, Will, "Progress in AI Isn't as Impressive as You Might Think," *MIT Technology Review*, November 30, 2017. As of August 4, 2020:
https://www.technologyreview.com/s/609611/
progress-in-ai-isnt-as-impressive-as-you-might-think/

Kravchenko, Stepan, "Future of Genetically Modified Babies May Lie in Putin's Hands," Bloomberg, September 29, 2019.

LaFleur, Karl, Kaitlin Cassady, Alexander Doud, Kaleb Shades, Eitan Rogin, and Bin He, "Quadcopter Control in Three-Dimensional Space Using a Noninvasive Motor Imagery-Based Brain–Computer Interface," *Journal of Neural Engineering,* Vol. 10, No. 4, 2013.

Lander, Eric S., Françoise Baylis, Feng Zhang, Emmanuelle Charpentier, Paul Berg, Catherine Bourgain, Bärbel Friedrich, J. Keith Joung, Jinsong Li, David Liu, et al., "Adopt a Moratorium on Heritable Genome Editing," *Nature,* Vol. 567, No. 7747, March 2019, pp. 165–168.

Lee, Mary, "The 'Internet of Bodies' Is Setting Dangerous Precedents," *Washington Post,* October 15, 2018.

Lee, Mary, Benjamin Boudreaux, Ritika Chaturvedi, Sasha Romanosky, and Bryce Downing, *The Internet of Bodies: Opportunities, Risks, and Governance,* Santa Monica, Calif.: RAND Corporation, RR-3226-RC, 2020. As of September 16, 2021:
https://www.rand.org/pubs/research_reports/RR3226.html

Leuthardt, Eric C., Jarod L. Roland, and Wilson Z. Ray, "Neuroprosthetics." *The Scientist,* October 31, 2014. As of August 4, 2020:
https://www.the-scientist.com/features/neuroprosthetics-36510

Levenson, Eric, "It Started as a Hobby. Now They're Using DNA to Help Cops Crack Cold Cases," CNN, March 27, 2019.

Levine, David I., "Automation as Part of the Solution," *Journal of Management Inquiry,* February 4, 2019. As of August 4, 2020:
https://journals.sagepub.com/doi/abs/10.1177/1056492619827375

Li Jianhua, "Coronavirus Pandemic Fuels China's Nucleic Acid Testing Industry," CGTN, February 22, 2021.

Li, Nan, Dave W. Oyler, Mengxuan Zhang, Yildiray Yildiz, Ilya Kolmanovsky, and Anouck R. Girard, "Game Theoretic Modeling of Driver and Vehicle Interactions for Verification and Validation of Autonomous Vehicle Control Systems," *IEEE Transactions on Control Systems Technology,* Vol. 26, No. 5, 2018, pp. 1782–1797.

Lilley, Kevin, "20,000 Soldiers Tapped for Army Fitness Program's 2nd Trial," *Army Times,* July 27, 2015.

Lindsey, Nicole, "Internet of Bodies: The Privacy and Security Implications," *CPO Magazine,* December 14, 2018.

Lopatto, Elizabeth, "Elon Musk Unveils Neuralink's Plans for Brain-Reading 'Threads' and a Robot to Insert Them," *The Verge,* July 16 2019. As of July 26, 2020:
https://www.theverge.com/2019/7/16/20697123/
elon-musk-neuralink-brain-reading-thread-robot

MacDonald, Marcy E., Christine M. Ambrose, Mabel P. Duyao, Richard H. Myers, Carol Lin, Lakshmi Srinidhi, Glenn Barnes, Sherryl A. Taylor, Marianne James, Nicolet Groat, et al., "A Novel Gene Containing a Trinucleotide Repeat That Is Expanded and Unstable on Huntington's Disease Chromosomes," *Cell,* Vol. 72, No. 6, 1993, pp. 971–983.

Maksimenko, Vladimir A., Alexander E. Hramov, Vadim V. Grubov, Vladimir O. Nedaivozov, Vladimir V. Makarov, and Alexander N. Pisarchik, "Nonlinear Effect of Biological Feedback on Brain Attentional State," *Nonlinear Dynamics,* Vol. 95, No. 3, 2019, pp. 1923–1939.

Manyika, James, and Kevin Sneader, "AI, Automation, and the Future of Work: Ten Things to Solve For," McKinsey Global Institute, June 1, 2018. As of August 4, 2020:
https://www.mckinsey.com/featured-insights/future-of-work/
ai-automation-and-the-future-of-work-ten-things-to-solve-for

Marks, Gene, "'Femtech' Startups on the Rise as Investors Scent Profits in Women's Health," *The Guardian,* June 6, 2019.

Marquet, Sandrine, "Overview of Human Genetic Susceptibility to Malaria: From Parasitemia Control to Severe Disease," *Infection, Genetics, and Evolution: Journal of Molecular Epidemiology and Evolutionary Genetics in Infectious Diseases,* Vol. 66, December 2018, pp. 399–409.

Marr, Bernard, "Meet The World's Most Valuable AI Startup: China's SenseTime," *Forbes,* June 17, 2019.

Marria, Vishal, "The Future of Artificial Intelligence in the Workplace," *Forbes,* January 11, 2019.

Marzbani, Hengameh, Hamid Reza Marateb, and Marjan Mansourian, "Neurofeedback: A Comprehensive Review on System Design, Methodology and Clinical Applications," *Basic and Clinical Neuroscience,* Vol. 7, No. 2, 2016.

Matwyshyn, Andrea M., homepage, undated. As of July 20, 2020:
https://www.andreamm.com/

Matwyshyn, Andrea M., "The 'Internet of Bodies' Is Here. Are Courts and Regulators Ready?" *Wall Street Journal,* November 12, 2018.

Mayo Clinic, Biofeedback, webpage, undated. As of May 6, 2020:
https://www.mayoclinic.org/tests-procedures/biofeedback/about/
pac-20384664

McCormack, Matthew P., and Terence H. Rabbitts, "Activation of the T-Cell Oncogene LMO2 After Gene Therapy for X-Linked Severe Combined Immunodeficiency," *New England Journal of Medicine*, Vol. 350, No. 9, 2004, pp. 913–922.

Mendell, Jerry R., Samiah Al-Zaidy, Richard Shell, W. Dave Arnold, Louise R. Rodino-Klapac, Thomas W. Prior, Linda Lowes, Lindsay Alfano, Katherine Berry, Kathleen Church, et al., "Single-Dose Gene-Replacement Therapy for Spinal Muscular Atrophy," *New England Journal of Medicine*, Vol. 377, No. 18, 2017, pp. 1713–1722.

Mesko, Bertalan, "The Role of Artificial Intelligence in Precision Medicine," *Expert Review of Precision Medicine and Drug Development*, Vol. 2, No. 5, September 2017, pp. 1–3.

Microsoft, "Democratizing AI in Health," webpage, undated. As of July 20, 2020:
http://democratizing-ai-in-health.azurewebsites.net/

Miliard, Mike, "Accenture, Merck Partner with Amazon Web Services for New Cloud Precision Medicine Platform," *Healthcare IT News*, September 17, 2018.

Mueller, Robert T., "New EEG Technology Makes for Better Brain Reading," *Psychology Today*, September 18, 2014.

Muse, sales webpage, undated. As of June 10, 2020:
https://choosemuse.com/shop/

National Academies of Sciences, Engineering, and Medicine, "International Commission on the Clinical Use of Human Germline Genome Editing," webpage, undated. As of February 23, 2020:
https://www.nationalacademies.org/gene-editing/international-commission/

National Academies of Sciences, Engineering, and Medicine, *Human Genome Editing: Science, Ethics, and Governance*, Washington, D.C.: National Academies Press, 2017.

National Human Genome Research Institute and European Molecular Biology Laboratory—European Bioinformatics Institute, GWAS Catalog, database, undated. As of June 13, 2019:
https://www.ebi.ac.uk/gwas/

National Research Council, *Emerging Cognitive Neuroscience and Related Technologies*, Washington, D.C.: National Academies Press, 2008.

Neal, Meghan, "The Internet of Bodies Is Coming, and You Could Get Hacked," *Motherboard*, March 13, 2014.

Nelson, Christopher E., Chady H. Hakim, David G. Ousterout, Pratiksha I. Thakore, Eirik A. Moreb, Ruth M. Castellanos Rivera, Sarina Madhavan, Xiufang Pan, F. Ann Ran, Winston X. Yan, et al., "In Vivo Genome Editing Improves Muscle Function in a Mouse Model of Duchenne Muscular Dystrophy," *Science*, Vol. 351, No. 6271, 2016, pp. 403–407.

Nelson, Patrick, "5G and 6G Wireless Technologies Have Security Issues," *Network World*, October 25, 2018.

Neuralink, homepage, undated. As of July 20, 2020:
https://www.neuralink.com

NeuroSky, "Enabling Technologies for Next-Generation mHealth Solutions," webpage, undated. As of June 12, 2020:
http://neurosky.com/about-neurosky/

NeuroTechEdu, "Consumer EEG Headsets," webpage, undated. As of July 12, 2020:
http://learn.neurotechedu.com/headsets/

Ng, Alfred, "Security Flaw Allows for Spying over 5G, Researchers Warn," NET, February 1, 2019.

Nicol, David M., "Hacking the Lights Out," *Scientific American*, Vol. 305, No. 1, 2011, pp. 70–75.

Nobel Foundation, "The Nobel Prize in Physiology or Medicine 2019," webpage, 2019. As of July 20, 2020:
https://www.nobelprize.org/prizes/medicine/2019/summary/

Normile, Dennis, "China Sprints Ahead in CRISPR Therapy Race," *Science*, Vol. 358, No. 6359, 2017, pp. 20–21.

Novembre, John, Toby Johnson, Katarzyna Bryc, Zoltán Kutalik, Adam R. Boyko, Adam Auton, Amit Indap, Karen S. King, Sven Bergmann, Matthew R. Nelson, et al., "Genes Mirror Geography Within Europe," *Nature*, Vol. 456, No. 7218, 2008, pp. 98–101.

Oberhaus, Daniel, "This DIY Implant Lets You Stream Movies from Inside Your Leg," *Wired*, August 30, 2019.

Office of Science and Technology Policy, Summary of the 2019 White House Summit on America's Bioeconomy, White House, October 2019.

O'Malley, Brendan, "China Is 'Systematically Stealing US Research'—Senate," *University World News*, November 22, 2019. As of February 22, 2020:
https://www.universityworldnews.com/post.php?story=20191122145800927

O'Meara, Sarah, "Will China Lead the World in AI by 2030?" *Nature*, Vol. 572, No. 7770, August 2018, pp. 427–428.

Online Mendelian Inheritance in Man, "OMIM Gene Map Statistics," webpage, May 4, 2020. As of May 5, 2020:
https://www.omim.org/statistics/geneMap

OpenBCI, "Shipping and Taxes," webpage, April 11, 2020. As of July 12, 2020:
https://docs.openbci.com/docs/08FAQ/
ShippingFAQ#what-are-the-shipping-rates

Osmakova, Alina, Michael Kirpichnikov, and Vladimir Popov, "Recent Biotechnology Developments and Trends in the Russian Federation," *New Biotechnology*, Vol. 40, Part A, January 25, 2019, pp. 76–81.

Osoba, Osonde A., and William Welser IV, *An Intelligence in Our Image: The Risk of Bias and Errors in Artificial Intelligence*, Santa Monica, Calif.: RAND Corporation, PE-237-RC, 2017. As of August 4, 2020:
https://www.rand.org/pubs/perspectives/PE237.html

O'Sullivan, Shane, Nathalie Nevejans, Colin Allen, Andrew Blyth, Simon Leonard, Ugo Pagallo, Katharina Holzinger, Andreas Holzinger, Mohammed Imran Sajid, and Hutan Ashrafian, "Legal, Regulatory, and Ethical Frameworks for Development of Standards in Artificial Intelligence (AI) and Autonomous Robotic Surgery," *International Journal of Medical Robotics and Computer Assisted Surgery*, Vol. 15, No. 1, February 2019.

Peake, Jonathan M., Graham Kerr, and John P. Sullivan, "A Critical Review of Consumer Wearables, Mobile Applications, and Equipment for Providing Biofeedback, Monitoring Stress, and Sleep in Physically Active Populations," *Frontiers in Physiology*, Vol. 9, June 2018, p. 743.

Pearson, Sue, Hepeng Jia, and Keiko Kandachi, "China Approves First Gene Therapy," *Nature Biotechnology*, Vol. 22, No. 1, 2004, pp. 3–4.

Philippidis, Alex, "10 Countries in 100k Genome Club," *Clinical OMICs*, August 30, 2018. As of July 12, 2019:
https://www.clinicalomics.com/topics/biomarkers-topic/
biobanking/10-countries-in-100k-genome-club/

Polderman, Tinca J. C., Beben Benyamin, Christiaan A de Leeuw, Patrick F Sullivan, Arjen van Bochoven, Peter M Visscher, and Danielle Posthuma, "Meta-Analysis of the Heritability of Human Traits Based on Fifty Years of Twin Studies," *Nature Genetics*, Vol. 47, No. 7, 2015, pp. 702–709.

Polyakova, Alina, "Weapons of the Weak: Russia and AI-Driven Asymmetric Warfare," Brookings Institution, November 15, 2018. As of August 4, 2020:
https://www.brookings.edu/research/
weapons-of-the-weak-russia-and-ai-driven-asymmetric-warfare/

Pontin, Jason, "The Genetics (and Ethics) of Making Humans Fit for Mars," *Wired*, July 7, 2018. As of June 20, 2019:
https://www.wired.com/story/
ideas-jason-pontin-genetic-engineering-for-mars/

Pramanik, Pijush Kanti Dutta, Saurabh Pal, and Prasenjit Choudhury, "Beyond Automation: The Cognitive IoT. Artificial Intelligence Brings Sense to the Internet of Things," in Arun Kumar Sangaiah, Arunkumar Thangavelu, and Venkatesan Meenakshi Sundaram, eds., *Cognitive Computing for Big Data Systems over IoT: Frameworks, Tools and Applications*, Springer, 2017, pp. 1–37.

President's Council on Bioethics, *Beyond Therapy: Biotechnology and the Pursuit of Happiness*, Washington, D.C., October 2003.

Project Baseline, homepage, undated. As of July 20, 2020: https://www.projectbaseline.com/

Public Law 110-233, Genetic Information Nondiscrimination Act, 2008.

Qiang, Yi, Pietro Artoni, Kyung Jin Seo, Stanislav Culaclii, Victoria Hogan, Xuanyi Zhao, Yiding Zhong, Xun Han, Po-Min Wang, Yi-Kai Lo, et al., "Transparent Arrays of Bilayer-Nanomesh Microelectrodes for Simultaneous Electrophysiology and Two-Photon Imaging in the Brain," *Science Advances*, Vol. 4, No. 9, September 2018.

Quin, Amy, "Fraud Scandals Sap China's Dream of Becoming a Science Superpower," *New York Times*, October 13, 2017.

Radu, Sintia, "Russia and China's High-Tech Bet," *U.S. News and World Report*, November 29, 2019.

Ramadan, Rabie, and Athanasios Vasilakos, "Brain Computer Interface: Control Signals Review," *Neurocomputing*, Vol. 223, No. 5, 2017, pp. 26–44.

Rankinen, Tuomo, Noriyuki Fuku, Bernd Wolfarth, Guan Wang, Mark A. Sarzynski, Dmitry Alexeev, Ildus I. Ahmetov, Marcel R. Boulay, Pawel Cieszczyk, Nir Eynon, et al., "No Evidence of a Common DNA Variant Profile Specific to World Class Endurance Athletes," *PLoS One*, Vol. 11, No. 1, 2016.

Raza, Haider, Dheeraj Ratheeb, Shang-Ming Zhou, Hubert Cecotti, and Girijesh Prasad, "Covariate Shift Estimation Based Adaptive Ensemble Learning for Handling Non-Stationarity in Motor Imagery Related EEG-Based Brain-Computer Interface," *Neurocomputing*, Vol. 343, May 2019, pp. 154–166.

Reardon, Sara, "Gene Edits to 'CRISPR Babies' Might Have Shortened Their Life Expectancy," *Nature*, Vol. 570, No. 7759, June 2019, pp. 16–17.

Regalado, Antonio, "The DIY Designer Baby Project Funded with Bitcoin: Cryptocurrency, Biohacking, and the Fantastic Plan for Transgenic Humans," *MIT Technology Review*, February 1, 2019.

Reimer, Bryan, "Elon Musk and the Tesla Automation Strategy: A Disruptor in Vehicle Safety or Not?" *Forbes*, April 28, 2019.

Ribeil, Jean-Antoine, Salima Hacein-Bey-Abina, Emmanuel Payen, Alessandra Magnani, Michaela Semeraro, Elisa Magrin, Laure Caccavelli, Benedicte Neven, Philippe Bourget, Wassim El Nemer, et al., "Gene Therapy in a Patient with Sickle Cell Disease," *New England Journal of Medicine,* Vol. 376, No. 9, February 28, 2017, pp. 848–855.

Saba, Luca, Mainak Biswas, Venkatanareshbabu Kuppili, Elisa Cuadrado Godia, Harman S. Suri, Damodar Reddy Edla, Tomaž Omerzu, John R. Laird, Narendra N. Khanna, Sophie Mavrogeni, et al., "The Present and Future of Deep Learning in Radiology," *European Journal of Radiology,* Vol. 114, May 1, 2019, pp. 14–24.

Sanders, Sean, and Jackie Oberst, eds., *Sponsored Collection: Precision Medicine and Cancer Immunology in China*, supplement to *Science* magazine, Vol. 359, No. 6375, February 2, 2018.

Saracco, Roberto, "2019 Tech Trends," *IEEE Future Directions Tech Blog,* December 22, 2018. As of July 20, 2020:
https://cmte.ieee.org/futuredirections/2018/12/22/2019-tech-trends/

Sattler, Colin M., *Aviation Artificial Intelligence: How Will It Fare in the Multi-Domain Environment?* Fort Leavenworth, Kan.: School of Advanced Military Studies, U.S. Army Command and General Staff College, 2018.

Sauce, Bruno, and Louis D. Matzel, "The Paradox of Intelligence: Heritability and Malleability Coexist in Hidden Gene-Environment Interplay," *Psychological Bulletin*, Vol. 144, No. 1, 2018, pp. 26–47.

Savage, Maddy, "Thousands of Swedes Are Inserting Microchips Under Their Skin," NPR, October 22, 2018.

Schmid, Rolf D., and Xin Xiong, "Biotech in China 2021, at the Beginning of the 14th Five-Year Period ("145")," *Applied Microbiology and Biotechnology,* May 3, 2021, pp. 1–15.

Schmidt, Michael A., and Thomas J. Goodwin, "Personalized Medicine in Human Space Flight: Using Omics Based Analyses to Develop Individualized Countermeasures That Enhance Astronaut Safety and Performance," *Metabolomics*, Vol. 9, No. 6, 2013, pp. 1134–1156.

Scudellari, Megan, "DARPA Funds Ambitious Brain-Machine Interface Program," *IEEE Spectrum*, May 21, 2019. As of August 4, 2020:
https://spectrum.ieee.org/the-human-os/biomedical/bionics/darpa-funds-ambitious-neurotech-program

Seyhan, Attila A., and Claudio Carini, "Are Innovation and New Technologies in Precision Medicine Paving a New Era in Patients Centric Care?" *Journal of Translational Medicine,* Vol. 17, No. 114, April 5, 2019.

Shankland, Stephen, "Elon Musk Says Neuralink Plans 2020 Human Test of Brain-Computer Interface," CNET, July 17, 2019.

Shead, Sam, "The VC Arm of Google's Parent Company Is Betting Its Billions on Life-Enhancing Healthcare Startups," *Business Insider,* September 10, 2017.

Shobert, Benjamin, "Meet the Chinese Company That Wants to Be the Intel of Personalized Medicine," *Forbes,* January 18, 2017.

Siggens, Lee, and Karl Ekwall, "Epigenetics, Chromatin and Genome Organization: Recent Advances from the ENCODE Project," *Journal of Internal Medicine,* Vol. 276, No. 3, 2014, pp. 1645–1664.

Silva, Gabriel A., "A New Frontier: The Convergence of Nanotechnology, Brain Machine Interfaces, and Artificial Intelligence," *Frontiers in Neuroscience,* Vol. 12, 2018.

Silver, Mike, "Scientists Develop Tiny Tooth-Mounted Sensors That Can Track What You Eat," *Tufts Now,* March 22, 2018.

Smith, Roger, "How Robots and AI Are Creating the 21st-Century Surgeon," *Robotics Business Review,* February 7, 2019.

Snider, Jamie, Max Kotlyar, Punit Saraon, Zhong Yao, Igor Jurisica, and Igor Stagljar, "Fundamentals of Protein Interaction Network Mapping," *Molecular Systems Biology,* Vol. 11, No. 12, December 2015.

Sniekers, Suzanne, Sven Stringer, Kyoko Watanabe, Philip R. Jansen, Jonathan R. I. Coleman, Eva Krapohl, Erdogan Taskesen, Anke R. Hammerschlag, Aysu Okbay, Delilah Zabaneh, et al., "Genome-Wide Association Meta-Analysis of 78,308 Individuals Identifies New Loci and Genes Influencing Human Intelligence," *Nature Genetics,* Vol. 49, No. 7, 2017.

St. John, Mark, David A. Kobus, Jeffrey G. Morrison, and Dylan Schmorrow, "Overview of the DARPA Augmented Cognition Technical Integration Experiment," *International Journal of Human-Computer Interaction,* Vol. 17, No. 2, 2004, pp. 131–149.

"State AGs Call for Data Regulations," document reprint, *Washington Post,* June 24, 2019.

Statista, "Wearables," webpage, undated. As of September 5, 2019: https://www.statista.com/outlook/319/117/wearables/china

Streitz, Norbert, "Beyond "Smart-Only' Cities: Redefining the 'Smart-Everything' Paradigm," *Journal of Ambient Intelligence and Humanized Computing,* Vol. 10, No. 2, February 14, 2019, pp. 791–812.

Suchkov, Sergey, H. Abe, E. N. Antonova, P. Barach, B. T. Velichkovskiy, M. M. Galagudza, D. A. Dworaczyk, D. Dimmock, V. M. Zemskov, I. E. Koltunov, et al., "Personalized Medicine as an Updated Model of National Health-Care System, Part 1: Strategic Aspects of Infrastructure," *Rossiyskiy vestnik perinatologii i pediatrii [Russian Bulletin of Perinatology and Pediatrics],* Vol. 62, No. 3, 2017a, pp. 7–14.

Suchkov, Sergey, H. Abe, E. N. Antonova, P. Barach, B. T. Velichkovskiy, M. M. Galagudza, D. A. Dworaczyk, D. Dimmock, V. M. Zemskov, I. E. Koltunov, et al., "Personalized Medicine as an Updated Model of National Health-Care System, Part 2: Towards Public and Private Partnerships," *Rossiyskiy vestnik perinatologii i pediatrii [Russian Bulletin of Perinatology and Pediatrics]*, Vol. 62, No. 4, 2017b, pp. 12–18.

Texas Business and Commerce Code, Title 11, Subtitle A, Chapter 503, 2009.

Three Square Market, homepage, undated. As of July 20, 2020:
https://www.32market.com/public/

Tian, Chao, Bethann S. Hromatka, Amy K. Kiefer, Nicholas Eriksson, Suzanne M. Noble, Joyce Y. Tung, and David A. Hinds, "Genome-Wide Association and HLA Region Fine-Mapping Studies Identify Susceptibility Loci for Multiple Common Infections," *Nature Communications*, Vol. 8, No. 1, 2017.

Tkachenko, Stanislav, "Future Complicated for Russian Biotech," *Russia Beyond*, October 30, 2014.

Topol, Eric J., "High-Performance Medicine: The Convergence of Human and Artificial Intelligence," *Nature Medicine*, Vol. 25, No. 1, 2019, pp. 44–56.

Tucker, Patrick, "Tomorrow Soldier: How the Military Is Altering the Limits of Human Performance," *Defense One*, July 12, 2017. As of July 12, 2018:
https://www.defenseone.com/technology/2017/07/tomorrow-soldier-how-military-altering-limits-human-performance/139374/

Turek, Matt, "Explainable Artificial Intelligence (XAI)," DARPA program information, undated. As of July 20, 2020:
https://www.darpa.mil/program/explainable-artificial-intelligence

United States Government Policy for Institutional Oversight of Life Sciences Dual Use Research of Concern, September 24, 2014. As of February 22, 2020:
https://www.phe.gov/s3/dualuse/Documents/durc-policy.pdf

U.S.-China Economic and Security Review Commission, "Section 2: Emerging Technologies and Military-Civil Fusion: Artificial Intelligence, New Materials, and New Energy," in *2019 Report to Congress*, Washington, D.C.: U.S. Government Publishing Office, November 2019.

U.S. Department of Health and Human Services, "Health Information Privacy," webpage, undated. As of July 13, 2019:
https://www.hhs.gov/hipaa/index.html

U.S. Food and Drug Administration, *Digital Health Innovation Action Plan*, undated. As of July 20, 2020:
https://www.fda.gov/media/106331/download

U.S. Food and Drug Administration, "FDA Approval Brings First Gene Therapy to the United States," news release, August 30, 2017. As of July 23, 2020:
https://www.fda.gov/news-events/press-announcements/
fda-approval-brings-first-gene-therapy-united-states

U.S. Food and Drug Administration, "How to Determine If Your Product Is a Medical Device," webpage, December 16, 2019. As of July 20, 2020:
https://www.fda.gov/medical-devices/classify-your-medical-device/
how-determine-if-your-product-medical-device

U.S. National Library of Medicine, ClinicalTrials.gov, database, undated. As of July 12, 2020:
https://clinicaltrials.gov

UserManual.wiki, "Interaxon MU02 Bluetooth LE Device User Manual R1," webpage, undated. As of June 12, 2020:
https://usermanual.wiki/Interaxon/MU02/html

Vangay, Pajau, Benjamin M.Hillmann, and Dan Knights, "Microbiome Learning Repo (ML Repo): A Public Repository of Microbiome Regression and Classification Tasks," *GigaScience*, Vol. 8, No. 5, May 2018, p. giz042.

Vermont Public Law No. H.764, Act 171, An Act Relating to Data Brokers and Consumer Protection, 2019.

Vincent, James, "The State of AI in 2019," *The Verge,* January 28, 2019. As of August 4, 2020:
https://www.theverge.com/2019/1/28/18197520/
ai-artificial-intelligence-machine-learning-computational-science

VivoKey, homepage, undated. As of July 20, 2020:
https://www.vivokey.com/

Wang, Echo, "China's Kunlun Tech Agrees to U.S. Demand to Sell Grindr Gay Dating App," Reuters, May 13, 2019.

Wang, Tong-Min, Guo-Ping Shen, Ming-Yuan Chen, Jiangbo Zhang, Y. Sun, Jing He, Wen-Qiong Xue, Xi-Zhao Li, Shao-Yi Huang, Xiao-Hui Zheng, et al., "Genome-Wide Association Study of Susceptibility Loci for Radiation-Induced Brain Injury," *Journal of the National Cancer Institute*, Vol. 111, No. 6, 2018.

Watanabe, Tokiko, and Yoshihiro Kawaoka, "Influenza Virus-Host Interactomes as a Basis for Antiviral Drug Development," *Current Opinion in Virology,* Vol. 14, October 2015, pp. 71–78.

Wee, Sui-Lee, "China Uses DNA to Track Its People, with the Help of American Expertise," *New York Times*, February 21, 2019.

Weinbaum, Cortney, Eric Landree, Marjory S. Blumenthal, Tepring Piquado, Carlos Ignacio, and Gutierrez Gaviria, *Ethics in Scientific Research: An Examination of Ethical Principles and Emerging Topics*, Santa Monica, Calif.: RAND Corporation, RR-2912-IARPA, 2019. As of July 20, 2020: https://www.rand.org/pubs/research_reports/RR2912.html

Weiss, Haley, "Why You're Probably Getting a Microchip Implant Someday," *The Atlantic*, September 21, 2018.

West, Darrell M., "What Is Artificial Intelligence?" Brookings Institution, October 4, 2018. As of August 4, 2020: https://www.brookings.edu/research/what-is-artificial-intelligence/

West, Darrell M., and John R. Allen, "How Artificial Intelligence Is Transforming the World," Brookings Institution, April 24, 2018. As of August 4, 2020: https://www.brookings.edu/research/how-artificial-intelligence-is-transforming-the-world/

Westall, Sylvia, and Ivan Levingston, "Chinese Genetics Firm's Testing in Middle East Raises New U.S. Tensions," Bloomberg, May 20, 2020. As of June 10, 2020: https://www.bloomberg.com/news/articles/2020-05-20/chinese-genetics-firm-s-mideast-testing-raises-new-u-s-tensions

Wexler, Anna, "Separating Neuroethics from Neurohype," *Nature: Biotechnology*, Vol. 16, No. 9, September 2018, pp. 988–990.

Whalen, Jeanne, "Medtronic's Ireland Move Results in Lower Taxes," *Wall Street Journal*, September 28, 2015.

Willems, Sara M., Daniel J. Wright, Felix R. Day, Katerina Trajanoska, Peter K. Joshi, John A. Morris, Amy M. Matteini, Fleur C. Garton, Niels Grarup, Nikolay Oskolkov, et al., "Large-Scale GWAS Identifies Multiple Loci for Hand Grip Strength Providing Biological Insights into Muscular Fitness," *Nature Communications*, Vol. 8, No. 16015, 2017.

Wittes, Benjamin, and Jane Chong, *Our Cyborg Future: Law and Policy Implications*, Washington, D.C.: Brookings Institution, September 2014.

Wolpaw, Jonathan R., and Dennis J. McFarland, "Control of a Two-Dimensional Movement Signal by a Noninvasive Brain-Computer Interface in Humans," *Proceedings of the National Academy of Sciences*, Vol. 101, No. 51, pp. 17849-17854.

Wood, Valerie, Antonia Lock, Midori Harris, Kim Rutherford, Jürg Bähler, and Stephen G. Oliver, "Hidden in Plain Sight: What Remains to Be Discovered in the Eukaryotic Proteome?" *Open Biology*, Vol. 9, No. 2, 2019.

Woolfson, Derek N., Gail J. Bartlett, Antony J. Burton, Jack W. Heal, Ai Niitsu, Andrew R. Thomson, and Chris W. Wood, "De Novo Protein Design: How Do We Expand into the Universe of Possible Protein Structures?" *Current Opinion in Structural Biology*, Vol. 33, August 2015, pp. 16–26.

Wootson, Cleve R., "A Man Detailed His Escape from a Burning House. His Pacemaker Told Police a Different Story," *Washington Post*, February 8, 2017.

World Health Organization, "Can Personalized Medicine Contribute to Prevention and Control of NCDs in the Russian Federation?" press release, May 14, 2018. As of July 20, 2020:
http://www.euro.who.int/en/media-centre/events/events/2018/05/can-personalized-medicine-contribute-to-prevention-and-control-of-ncds-in-the-russian-federation

Xi, Lei, Jianfeng Chen, Yuehua Huang, Yanchun Xu, Lang Liu, and Yimin Zhou, "Smart Generation Control Based on Multi-Agent Reinforcement Learning with the Idea of the Time Tunnel," *Energy*, Vol. 153, Issue C, 2018, pp. 977–987.

Xue, Katherine S., Louise H. Moncla, Trevor Bedford, and Jesse D. Bloom, "Within-Host Evolution of Human Influenza Virus," *Trends in Microbiology*, Vol. 26, No. 9, 2018, pp. 781–793.

Yang, Luhan, Marc Güell, Dong Niu, Haydy George, Emal Lesha, Dennis Grishin, John Aach, Ellen Shrock, Weihong Xu, Jürgen Poci, et al., "Genome-Wide Inactivation of Porcine Endogenous Retroviruses (PERVs)," *Science*, Vol. 350, No. 6264, 2015, pp. 1101–1104.

Yudell, Michael, Dorothy Roberts, Rob DeSalle, and Sarah Tishkoff, "Science and Society: Taking Race out of Human Genetics," *Science*, Vol. 351, No. 627? 2016, pp. 564–565.